Developmental Neurobiology of Arthropods

Developmental Neurobiology of Arthropods

Edited by
D. YOUNG
Research Fellow in Neurobiology
Institute of Advanced Studies, Australian National University

Cambridge University Press

Published by the Syndics of the Cambridge University Press
Bentley House, 200 Euston Road, London NW1 2DB
American Branch: 32 East 57th Street, New York, N.Y.10022

Library of Congress Catalogue Card Number: 72–97883

ISBN: 0 521 20229 9

First published 1973

Printed in Great Britain
at the University Printing House, Cambridge
(Brooke Crutchley, University Printer)

Contents

Contributors

ALOE, L., Department of Biology, Washington University, St Louis, Missouri 63130, USA.

BATE, C. M., Department of Neurobiology, Research School of Biological Sciences, Australian National University, Canberra, ACT 2600, Australia.

BENTLEY, D. R., Department of Zoology, University of California, Berkeley, California 94720, USA.

CHEN, J. S., Laboratorio di Biologia Cellulare, CNR, Via G. Romagnosi 18/A, 00196 Rome, Italy.

EDWARDS, J. S., Department of Zoology, University of Washington, Seattle, Washington 98195, USA.

HORRIDGE, G. A., Department of Neurobiology, Research School of Biological Sciences, Australian National University, Canberra, ACT 2600, Australia.

HOY, R. R., Division of Biological Sciences, State University of New York, Stony Brook, NY 11790, USA.

LAWRENCE, P. A., MRC Laboratory of Molecular Biology, University Postgraduate Medical School, Hills Road, Cambridge CB2 2QH, England.

LEVI-MONTALCINI, R., Laboratorio di Biologia Cellulare, CNR, Via G. Romagnosi 18/A, 00196 Rome, Italy *and* Department of Biology, Washington University, St Louis, Missouri 63130, USA.

MEINERTZHAGEN, I. A., Department of Neurobiology, Research School of Biological Sciences, Australian National University, Canberra, ACT 2600, Australia.

PALKA, J., Department of Zoology, University of Washington, Seattle, Washington 98195, USA.

PIPA, R. L., Division of Entomology and Parasitology, University of California, Berkeley, California 94720, USA.

SESHAN, K. R., Department of Biology, Washington University, St Louis, Missouri 63130, USA.

YOUNG, D., Department of Neurobiology, Research School of Biological Sciences, Australian National University, Canberra, ACT 2600, Australia.

Editor's Introduction

D. Young

I believe that we should be concerned with the entire developmental process – that behaviour development will only be fully understood if we apply all the methods of experimental embryology to problems of behaviour development.
Tinbergen*

This statement by a distinguished ethologist effectively captures the spirit of the subject with which this book is concerned. For developmental neurobiology is compounded of developmental biology, which deals with the mechanisms of development from egg to adult, and neurobiology, which deals with the structure and function of nervous systems. The aim of developmental neurobiology is to seek an understanding of the mechanisms by which neuronal organisation is established during development, in relation to anatomy and physiology on the one hand and to ethology on the other. This subject has been studied on and off for about fifty years but the last few years have witnessed a widespread renewal of interest. This fresh interest may be attributed partly to techniques such as electron microscopy and the use of the microelectrode, which have enabled us to make a fresh attack on problems that have long been evident. Also it is attributable to the increasing recognition that the nervous system requires analysis in terms of intercellular communication and pattern formation at the level of the whole organism. Recent interest in these problems, which are not peculiar to the nervous system, is in contrast to the earlier concentration on the analysis of the simpler cell activities in biophysical and biochemical terms.

This renewed interest is reflected in a number of books which have appeared recently, reviewing the subject from various points of view. The work of Hughes, *Aspects of Neural Ontogeny* (Logos Press, 1968), was followed by the comprehensive review of Jacobson in *Developmental Neurobiology* (Holt, Rinehart & Winston, 1970) and the more restricted and intensive discussion by Gaze in

* This sentence is taken from a valuable outline of the history of ethology by Professor Tinbergen. The full reference is Tinbergen, N. (1969) Ethology. In *Scientific Thought 1900–1960*, ed. R. Harré, pp. 238–68. Oxford: Clarendon Press.

The Formation of Nerve Connections (Academic Press, 1970). There is also a collection of papers with more emphasis on the human nervous system, edited by Himwich under the title *Developmental Neurobiology* (Charles C. Thomas, 1970). These books concentrate almost entirely on vertebrate material, which is understandable because work on invertebrates is not voluminous and most of it is quite recent. In the present book, I have drawn together reviews by a number of people who are actively studying invertebrate, and especially arthropod, nervous systems from a developmental point of view.

The contribution which the studies on arthropods can make to the field may be viewed in two ways. On the one hand, the arthropods deserve study in their own right as one of the major animal phyla (and one which in its numerical strength reduces the vertebrates to insignificance!). From this zoological standpoint, the establishment of known similarities and differences in the mechanisms of the various invertebrate and vertebrate groups may be expected to contribute important insights to our understanding of neuronal development in all animals. On the other hand, arthropods are proving to be advantageous material for the study of general problems of developmental neurobiology. But here we should be cautious of treating the arthropod nervous system as a model system in relation to the vertebrates, for the arthropods are not simply impoverished vertebrates. Most of the contributors to this book would probably admit to a mixture of these points of view in their own work.

Some of the advantageous features of arthropods for this work are illustrated by the contributions which follow. One of these useful features is that the sensory neuron cell bodies are situated peripherally. Consequently, when a peripheral sensory structure is transplanted, or otherwise manipulated, the sensory neurons go with it and then regrow to the central nervous system. The subsequent interpretation of this regrowth is thus potentially less ambiguous than in the corresponding situation in vertebrates. This feature has been exploited by J. S. Edwards and J. Palka, whose contribution gives an account of specificity and regeneration in an insect sensory appendage, following various experimental alterations to the normal situation. Similarly C. M. Bate and P. A. Lawrence have employed this feature of the insect sensory nervous system in their experimental study and discussion of the specificity of cuticular sensory hairs in relation to the theory of gradients in the specification of insect cuticular patterns.

Another useful feature of the arthropod nervous system is that single neurons can be identified from one animal to another. The advantages of working with single identified motor neurons is apparent from D. R. Bentley's account of the development of the motor systems underlying flight and song behaviour patterns in crickets and locusts. My own contribution illustrates how single identified motor neurons are beginning to be used in the study of the specificity of nerve–muscle connections. A comparison of this topic within the arthropods is made possible by R. R. Hoy's review of motor nerve degeneration and regeneration in Crustacea.

Visual systems have been a favourite object of study among students of neuronal specificity and development, and the insect visual system has proved useful in this direction, as shown by I. A. Meinertzhagen's review of specificity and development in optic lobe neurons. But all this concentration of study at the cellular level should not lead us to overlook the larger scale changes which occur during the postembryonic development of the nervous system. These major changes in insects are reviewed by R. L. Pipa. It would be a further advantage if some of these topics could be pursued under *in-vitro* experimental conditions. That this can be done is now known through the work of R. Levi-Montalcini and her associates and they have summarised this promising development in their contribution. Finally, our book is flavoured with a speculative essay on specificity and development of neurons by G. A. Horridge which contains some interesting and controversial thoughts.

These are just a few sample points which I have picked out for attention from the following nine contributions and they are intended only to prepare the way briefly for study of the individual papers. No concluding summation is provided or any final drawing together of the threads of discussion. This is a subject which is changing fast and many contributors make mention of their expectation of rapid advance from our present position, with new data and new ideas developing rapidly. In this situation it seems wiser to leave the various threads of discussion loose, rather than to attempt premature generalisation and synthesis. But by each reviewing the situation as it appears to us in the early 1970s, we hope to have provided something of a focal point for future work

D. Young

in arthropod developmental neurobiology and also a useful compilation for interested workers and students in other, related fields of study.

I am most indebted to my wife, Rosemary, and to Mrs Carmen Vilcins, Mrs Yvonne Mortimore and Mrs Suzanne Tobin for their invaluable assistance in the editorial processing of this book.

An *in-vitro* approach to the insect nervous system

R. Levi-Montalcini, J. S. Chen, K. R. Seshan and L. Aloe

Introduction

To the 1972 neurobiologist the insect nervous system is a far more attractive object of investigation than it was to his predecessors, when the concept prevailed that insects and other invertebrates were too low in the phylogenetic scale to give meaningful information on the same system in vertebrates.

The invertebrates have often been looked at with curiosity and with the suspicion that they represent really an inferior brand of living organism, the lower end, as it were, of the evolutionary ladder which in the mind of the mid-nineteenth century biologists and their faithful pupils assumed the significance of a scale of values. What we are witnessing these days is the liberation from this scale of values – or at least a rapid ascendancy of the invertebrates on the very same scale. (Florey, 1967).

Evidence for this trend referred to by Florey is to be found in the ever growing list of studies on insects performed with the most exacting techniques, as exemplified by the articles in this same volume.

While the investigations at the structural, ultrastructural, electrophysiological and behavioural levels on the intact insect nervous system, or on parts of it, have progressed at a remarkably fast pace and have, in fact, moved ahead of the same studies on the vertebrate nervous system, at least in some areas, almost no attempts have been made to submit the insect nervous system to an *in-vitro* analysis.

The belief that these techniques, which proved to be so valuable in the study of the vertebrate nervous system, should likewise bring information on the insect nervous system, encouraged us to start four years ago a programme having as its object the nervous system of the cockroach, *Periplaneta americana*.

The choice proved to be particularly fortunate. It soon became apparent that the embryonic nervous system of this insect adapts remarkably well to the conditions of culture. As we became more and more familiar with the object of our studies, we realised the possibility of extending the investigations to new areas and kept probing a field which enlarged as we advanced in it. We feel that we have at present only barely tapped the source of information which became available by this *in-vitro* approach. For this reason, the investigations performed in these years are still of an exploratory nature and the results should be considered more for their prospective than for their present value. It is in keeping with this approach that we shall present a panoramic view of the field, rather than enlarging in detail on the results obtained in each of the many sectors of the nervous system which have been submitted to an *in-vitro* analysis. The studies to be reported in the following sections started in the spring of 1968 at Washington University and are now being continued in parallel at that University and in the Laboratory of Cell Biology of the CNR in Rome.

The object of choice: *Periplaneta americana*

The large size of this insect, its availability and the facility of rearing it in the laboratory are the main reasons why the cockroach *Periplaneta americana* became a favourite object of morphological, biochemical and electrophysiological studies. The same cockroach and closely related species, such as *Leucophaea maderae*, *Blabera fusca* and *Blatella germanica*, had been selected for organ and cell culture shortly before we became interested in the same problem (Landureau, 1966; Marks & Reinecke, 1965; Marks, Reinecke & Leopold, 1968). The results of these investigations will be considered in following sections.

Periplaneta americana seemed to us to offer several advantages over other blattoid species and was therefore selected for the *in-vivo* and *in-vitro* studies which are carried out in parallel in our laboratories on the embryonic, nymphal and adult insect. Here only the results of the experiments *in vitro* will be considered.

Each ootheca contains fourteen to sixteen embryos, thus affording the possibility of performing multiple experiments on brain and ganglia as well as other organs of specimens at the same developmental stage. For experiments *in vitro*, the sterility procedures are reduced to a minimum as compared to those required for single-embryo eggs or for viviparous forms. When compared with other insects, the cockroach offers the advantage of undergoing a progressive and step-wise increase of the whole organism with no significant changes in the general organisation of nerve structures, which can be studied from early embryonic development to maturity as a continuous process, thus offering the possibility of studying the same nerve cell populations during different stages of their entire cycle *in vivo* and *in vitro*. Finally, the long life span and slow developmental processes, which would bar the use of this insect for genetic studies, are additional assets and reasons for its selection for neurological purposes, since the sequence of developmental events is obviously easier to analyse when these processes take place in days and weeks than when they unfold in hours as is the case with the favourite object of genetics, the fruit fly.

The culture medium

We reported in previous articles on the vigorous nerve fibre outgrowth and cell migration from intact explants of brain, ganglia and other tissues from 16- to 18-day embryos, cultured in a medium consisting of 4 parts of the Eagle basal medium and 5 parts of the Schneider's *Drosophila* solution (Chen & Levi-Montalcini, 1969, 1970*a*; Levi-Montalcini & Chen, 1969). This medium, which in our hands and for the present purposes gave much better results than the Schneider or Grace medium alone or supplemented with calf or horse serum, crustacean haemolymph, extract of cockroach embryos, was developed through an entirely empirical procedure based on the screening of all media available in the market, including the vertebrate media. This trial and error technique, which may surprise the reader and convey the impression of too dilettante an approach, was, however, suggested by a comparative study of media developed in other laboratories and reported as satisfactory by different workers. The composition of some of these media which now find wide application in insect cultures, particularly of Diptera, is given in Table 1. It will be seen that they differ to such an extent in all the essential components, salts, sugars, organic acids, vitamins, amino acids, as to give no clue to the nutritional requirements of insect cells and to justify a 'shot in the dark' approach, at least at an early stage of our own project.

After months of unsuccessful and frustrating attempts, this empirical method paid off far beyond our expectations. The '4+5' medium, as we referred to it, was developed in the spring

Table 1. *Composition of media widely used in insect culture* (mg l^{-1})

	Basal (Eagle) – Schneider's *Drosophila* medium[1]	Schneider's *Drosophila* medium[2]	Grace's insect T.C. medium[3]	Landureau's medium S 19[4]	Mark's medium M 14[5]
INORGANIC SALTS					
NaCl	4189	2100	—	8500	1500
KCl	1067	1600	2240	1050	400
$NaH_2PO_4 . 10H_2O$	135	—	2202	—	400
$NaH_2PO_3 . H_2O$	—	—	—	—	200
KH_2PO_4	250	450	—	—	—
Na_2HPO_4	389	700	—	—	—
$MgSO_4 . 7H_2O$	2144	3700	2780	1260	—
$MgCl_2 . 6H_2O$	—	—	2280	—	100
$NaHCO_3$	1200	400	350	360	200
$CaCl_2$	422	600	1000	490	450
$Na_2SO_4 . 10H_2O$	—	—	—	—	100
CH_3COONa	—	—	—	—	100
NaOH	—	—	—	—	—
$MnSO_4 . H_2O$	—	—	—	65	—
PO_3H_3	—	—	—	900	—
SUGARS					
Glucose	1555	2000	700	3000	15000
Trehalose	1111	2000	—	—	3000
Fructose	—	—	400	—	10000
Sucrose	—	—	26680	—	5000
ORGANIC ACIDS AND VITAMINS					
α-Ketoglutaric acid	111	200	370	—	300
Fumaric acid	56	100	55	—	100
Malic acid	56	100	670	—	500
Succinic acid	56	100	60	—	200
Folic acid	0·44	—	0·02	0·01	0·02
Citric acid	—	—	—	—	100
Ascorbic acid	—	—	—	—	200
Biotin	0·44	—	0·010	0·01	0·01
Riboflavin	0·444	—	0·02	0·05	0·02
Choline Cl	0·44	—	0·02	0·4	0·02
Niacin	—	—	0·02	—	0·02
Nicotinamide	0·44	—	—	0·03	—
Ca-pantothenate	0·44	—	0·02	0·1	0·02
Pyridoxine HCl	—	—	0·02	0·03	0·02
Pyridoxal HCl	0·44	—	—	—	—
Thiamine HCl	0·44	—	0·02	0·01	0·02
L-Inositol	0·8	—	0·02	0·05	0·02
Carnitine	—	—	—	—	0·01
p-Aminobenzoic acid	—	—	0·02	—	0·02

Table 1 (*cont.*)

	Basal (Eagle) – Schneider's *Drosophila* medium[1]	Schneider's *Drosophila* medium[2]	Grace's insect T.C. medium[3]	Landureau's medium S 19[4]	Mark's medium M 14[5]
Amino acids					
l-Tyrosine	285	500	50	180	83
l-Cystine	61	100	22	—	87
l-Lysine HCl	930	1650	625	160	200
l-Methionine	448	800	50	500	300
l-Threonine	205	350	175	200	100
l-Leucine	94	150	75	250	200
l-Isoleucine	94	150	50	120	200
l-Arginine	231	400	580	800	496
β-Alanine	278	500	200	—	300
l-Aspartic acid	222	400	350	—	400
l-Cysteine	33	60	—	260	—
l-Histidine	225	400	2500	300	485
Glycine	139	250	650	750	300
l-Phenylalanine	91	150	150	200	200
l-Proline	944	1700	350	750	500
l-Serine	139	250	1100	80	200
l-Tryptophan	58	100	100	200	50
l-Valine	178	300	100	150	300
l-Glutamic acid	444	800	600	1500	1000
l-Glutamine	1130	1800	600	300	1000
l-Asparagine	—	—	350	250	600
α-Alanine	—	—	225	120	100
Taurine	—	—	—	—	100
Other components					
DPN-Nicotinamide-adenine dinucleotide	—	—	—	—	100
Glutathione	—	—	—	—	400
Glycyl-DL aspartic acid	—	—	—	—	200
Leucyl-DL glycyl-DL phenylalanine	—	—	—	—	200
Glycyl-glycyl-glycyl-glycine	—	—	—	—	200
Polyvinylpyrrolidone (PVP K90)	—	—	—	—	10000
Polyvinylpyrrolidone (PVP K30)	—	—	—	—	10000
Yeast hydrolysate	1111	2000	—	—	—
Phenol red	4·4	—	—	—	—

[1] Chen & Levi-Montalcini, 1969. [2] Schneider, 1964.
[3] Grace, 1962. [4] Landureau, 1969; Landureau & Jolles, 1969.
[5] Marks, Reinecke & Caldwell, 1968.

of 1968 and has been used ever since in our and other laboratories (Schlapfer, Haywood & Barondes, 1972) to grow embryonic, nymphal and adult nervous system of this insect, as well as intact organs or dissociated nerve cells, for periods of many months. At the time of the discontinuation of the cultures, most of them are in such excellent condition as to convey the impression that they could remain alive for much longer periods and perhaps indefinitely, provided care is taken to replace the medium every few days and to prevent bacterial or mould contamination. This is assured by adding 100 ml^{-1} penicillin, 100 $\mu g\ ml^{-1}$ streptomycin and 0·25 $\mu g\ ml^{-1}$ fungizone.

Until very recently we purchased both Schneider and Eagle media, mixing them in the above ratio at the moment of use. The frequent occurrence of salt precipitates in cultures suggested the preparation of the medium in our laboratory according to the formula given in Table 1. Yeast hydrolysate, which is present in the Schneider solution as indicated, proved not to be essential for nerve fibre outgrowth and cell migration from intact explants or for growth of dissociated nerve cells and is now omitted from the preparation which consists, therefore, of an entirely defined chemical medium.

However, the statement that this medium satisfies the nutritional requirements of embryonic and differentiated insect nerve tissues needs some qualification. Mitotic activity in the migratory area around non-nervous explants, such as segments of the alimentary tract, is very low; only one, or at best two, mitotic figures are seen in each two to three-week old culture. Dividing cells are very seldom found in fibroblast-like cells migrated out from these explants in cultures one to three months old. The situation seems to be even worse in leg regenerates cultured in the medium devised by Marks and his co-workers. The authors stated in fact that in these cultures 'only one mitotic event was recorded during four years' (Marks, Reinecke & Caldwell, 1968, p. 89).

While such low mitotic activity obviously reflects nutritional deficiencies in our medium which would prevent its use for the study of other cell lines, it is of much less consequence in the case of the nerve tissue, which does not undergo proliferation in this or in any other medium. Since nerve cells in our cultures not only produce profuse nerve fibre outgrowth and undergo size increase, but also build typical synapses between axons and between these and their end organs, it seems justified to continue use of this medium, while making intensive efforts to improve it. The systematic and most rigorous studies by Landureau on the amino acid requirements of blattoid cells (Landureau, 1969; Landureau & Jollès, 1969) are being utilised in our laboratory to improve this medium. This author succeeded in fact in culturing and sub-culturing different cell lines from *Periplaneta americana* for periods up to two years in the medium reported in Table 1 which, in turn, owes much to the studies by T. D. C. Grace on nutritional requirements of lepidopteran and dipteran cells and to his pioneer work in designing media for culturing cell lines of these insects (Grace, 1962, 1966).

A few words may also be said on the vastly different results obtained in our laboratory in cultures of single or multiple explants as well as of explants cultured in the presence of dissociated embryonic nerve cells. Although this aspect of our cultures will be considered again in the following sections, it is of interest to mention these differences in connection with the problem of nutritional requirements of nerve and non-nerve insect cells *in vitro*. We found in fact that explants of intact brain, ganglia, segments of the alimentary tract from cockroach embryos, as well as ganglia of the stomatogastric and neuro-endocrine systems of nymphs and adult insects, undergo progressive deterioration when cultured alone, whereas they survive in excellent condition when cultured in presence of other like or unlike embryonic explants or dissociated embryonic nerve cells. A most plausible explanation, and one also suggested by Landureau on the basis of his own similar experience with cultures of pure cell lines (Landureau, 1969; Landureau & Jollès, 1969), is that embryonic cells release into the medium aspecific factors which compensate for amino acids or other nutritional deficiencies in the culture media. Thus while these factors, released by growing cells, frustrate our attempts to devise an entirely defined medium which would support growth and differentiation of nerve tissues, they correct the incompleteness of synthetic media and provide an explanation for the remarkable ability of embryonic, as well as fully differentiated neurons, to spin out axons and establish connections through the agency of these fibres with other tissues, under conditions which depart to such an extent from those provided by the living organism. The capacity of insect nerve tissue to adjust to the culture medium used in our laboratories seems in fact to be far superior to that of vertebrate nerve tissue, which needs to be dissected in minute fragments not larger than 1 mm (Lumsden, 1968) before its explantation *in vitro* and, even so, lends itself to an *in-vitro* analysis to a much more limited extent than the insect nervous system. Insect brain and ganglia, while producing nerve fibres and cell migration *in vitro*, still retain much of their inner organisation even in long-term cultures. In the following sections we shall consider the main results of our analysis of the embryonic, nymphal and adult insect nervous system, which is presented more extensively in previous articles, as indicated.

The embryonic insect nervous system

The nervous system of a 16-day embryo is a miniature, but otherwise faithful, replica of that of the adult, at least at the macroscopic level; the marked difference in the overall size results from the diminutive size of nerve cells at this early developmental stage, rather than from differences in nerve cell number. Although in fact sporadic and scattered mitotic activity occurs among globulus cells of corpora pedunculata during all embryonic stages as well as in the nymph, the total number of nerve cells does not apparently undergo increase, because while some cells undergo division in late embryonic and larval stages, others undergo degeneration and are reabsorbed (Edwards, 1969). In 16-day embryos, the immature neurons which make up the brain and ganglion nerve cell populations, though small in size, have already acquired distinctive features which permit them to be distinguished from glial cells, such as the round shape and the thin cytoplasmic area which surrounds a large nucleus and bulging nucleolus. They are densely crowded in the peripheral ring around the neuropile in the three brain vesicles: proto-, deutero- and tritocerebrum. The neuropile, as shown in Plate 1*a* which reproduces a frontal section of a 16-day embryo stained with a specific silver technique, occupies a very large area. A comparison of Plate 1*a* and *b*, however, gives evidence for the marked difference in the structural organisation of cells and fibre tracts in the embryonic and adult brain. Thoracic and abdominal ganglia are likewise already differentiated in 16-day embryos and differ from the same ganglia of nymphal and adult insects in the size of neurons rather than in their number. Nerve

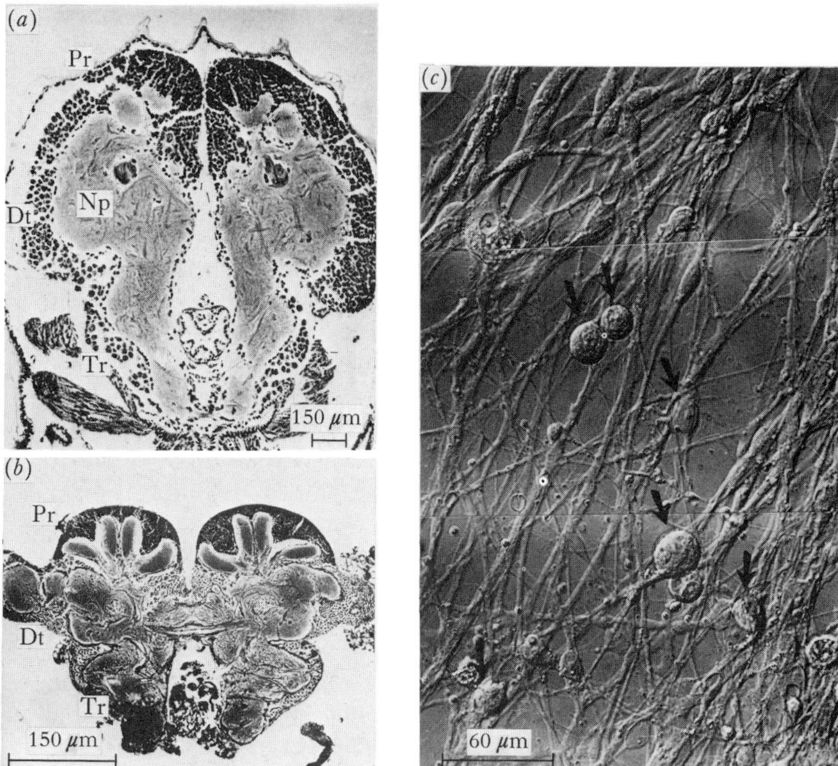

Plate 1. (*a*), (*b*) Frontal sections of brains of a 16-day embryo (*a*) and of an adult insect (*b*). Pr, protocerebrum; Dt, deutocerebrum; Tr, tritocerebrum; Np, neuropile. Chen modification of Bodian stain.

(*c*) Photomicrograph (Nomarski) of segment of migratory area around thoracic ganglion from a 16-day embryo cultured *in vitro* for 2 weeks and photographed in living culture. Arrows point to nerve cells migrated out from explant. Nerve fibres and spindle-shaped cells are also apparent.

fibres emerging from interneurons interconnect all ganglia in the ventral chain, while other fibres emerging from the large motor pool in the thoracic ganglia branch out and assemble in slender nerves which connect with the leg muscles. The digestive tract is well differentiated and the ingluvial ganglion adhering to the oesophagus is recognisable at the dissecting microscope. Both the ingluvial and the frontal ganglion, the two main components of the stomatogastric system, have already attained their full quota of neurons, as ascertained by cell counts performed in embryos and adult insects of our collection. The neurons in both ganglia are aligned in two cell rows around a central, well-developed neuropile.

Observations on living and fixed cultures of brain and ganglia

Embryos 16 and 18 days old were used in preference to younger and older embryos for these *in-vitro* studies. The technique for dissecting out brain, ganglia segments of the digestive tract, leg primordia, heart and other tissues, is given in previous publications (Chen & Levi-Montalcini, 1969, 1970*b*; Levi-Montalcini & Chen, 1969). Here it suffices to mention that nerve and non-nerve tissues, dissected out from the embryos, are cultured on cover slides laid on the bottom of small, cylindrical vessels (internal diameter, 13 mm; height, 7 mm), half filled with the culture medium. The vessels are then placed in desiccators flushed with 5% CO_2 in 95% air and are incubated at 29° in a vapour saturated atmosphere.

Observations on cultures were performed daily at the inverted microscope and with the Nomarski interference microscope; contractility of foregut segments, as well as of myocytes migrated out from this organ, was recorded with a cine-camera, as indicated in a previous article (Aloe & Levi-Montalcini, 1972*a*). Upon discontinuation of the experiments, the cultures were fixed for routine histological studies or, more frequently, for studies of nerve cells and fibres stained with the Cajal–De Castro technique, as already reported (Levi-Montalcini, 1963). Many cultures were also fixed and examined at the electron microscope. The culture medium was replaced at first every week and then, in older cultures, every three days.

The migrating nerve cells

Cells with unmistakable features of nerve cells are seen between the end of the first and subsequent weeks of culture around brain and ganglionic explants. They are much more numerous around thoracic than abdominal ganglia; a situation intermediate between the two is represented by brain explants, while the same cells are absent in the migratory zone around frontal and ingluvial ganglia. The criteria for the identification of these cells as nerve cells are listed in previous articles and need not be repeated here (Chen &

Levi-Montalcini, 1970*b*; Levi-Montalcini & Chen, 1969). In living cultures, the cells appear well defined against the background on account of their dense texture which comes into sharp relief at the Nomarski interference microscope, as shown in Plate 1*c*. Also characteristic of these cells are the smooth oval or round contour, the large nucleus and bulging nucleolus. At the electron microscope, regularly spaced microtubules are seen in the cell perikaryon, as well as in the two slender fibres identified as the cell axons, which emerge from the opposite ends of the cell. In silver stained preparations, the small round bodies of these cells are seen adhering to the surface of cells of very large size with distinct glial features. This most unusual cell-to-cell contact by apposition of the body of one cell to that of another, rather than juxtaposition of contiguous cells, finds no parallel in vertebrate cultures, where 'contact inhibition' prevents migrating cells from piling on other like or unlike cells. The only well known exception to this rule is that of neoplastic cells and, in fact, lack of contact inhibition has acquired significance as a revealing sign of deviation from normality (Abercrombie, 1958). In the present cultures, the situation is obviously different and does not indicate deviation from normality but rather is suggestive of the dependence of nerve cells on their ancillary cells, the glial cells; in the living organism the latter are tightly wrapped around the nerve cell perikarya and their axons. It is therefore conceivable that even in the different conditions of the culture, the close bonds which link neurons with glial cells still hold and account for this most singular type of cell-to-cell contact.

The question of whether nerve cells are endowed with the property of active motion or whether they are passively dragged out from the explant by glial cells is still not definitely settled, although many arguments presented in previous articles favour the first alternative. The finding that nerve cells are present in much larger numbers around thoracic ganglia than around brain and abdominal ganglia suggests that these neurons may be motor neurons rather than interneurons; the former are in fact much more numerous in thoracic ganglia than in other sectors of the insect nervous system, where interneurons prevail. Nerve cells decrease progressively in the migratory areas of older embryos and are not seen around explants of nerve tissue of nymphs and adult insects; thus only immature neurons would be able to move out – in one way or the other – from their site of origin.

The glial cells

Of the three glial cell types described around brain and ganglion explants in our cultures (Levi-Montalcini & Chen, 1969) we shall consider here only the most common one, which exhibits such distinctive glial cell characteristics as to leave no doubt as to its nature. Four features mark off these cells from others, namely their exceptionally large size, the flattened and thin cytoplasmic layer, the indented and broken contour, and the fibrillar network sculptured on their surface by nerve fibres adhering so firmly to the cell body as to give the impression that they are part of the same cells. The adhesion to their surface of cells identified as neurons was mentioned in the previous section. The broken and indented cell contours, as well as the frequent occurrence of cytoplasmic fragments torn away from the cell bodies and adhering to adjacent nerve fibres (Chen & Levi-Montalcini, 1969, 1970*b*; Levi-Montalcini & Chen, 1969), speak for an unusual fragility of these cells. Additional evidence for their frailty came from experiments reported on p. 22.

The nerve fibres

Nerve fibres emerge from the entire surface of intact brain and ganglia from 16-day embryos in the third day of incubation and increase progressively in number and length during subsequent weeks. The study of hundreds of explants of brain and ganglia, cultured for periods varying from a few weeks to many months, gave evidence for two distinct patterns of nerve fibre outgrowth from brain. thoracic and abdominal ganglia. The fibres emerge either as individual filaments (Plate 3*a*) or assembled in large bundles, which at their distal ends split into single units of such tenuous texture and diminutive size as to be barely visible at the light microscope in living or fixed preparations. Two of these colossal bundles are shown in Plate 2*a* and *b* which reproduce, respectively, a fibre bundle emerging from the protocerebrum and one from the sixth terminal abdominal ganglion in 20-day old cultures. The strong axon-to-axon affinity of fibres assembled in bundles is also indicated by their tendency to aggregate again after splitting, and in the fine cobweb which interconnects the thin threads as they branch, fan-like, into the medium. The two patterns of nerve fibre outgrowth described above are characteristic of different sectors of the embryonic nervous system; fibres

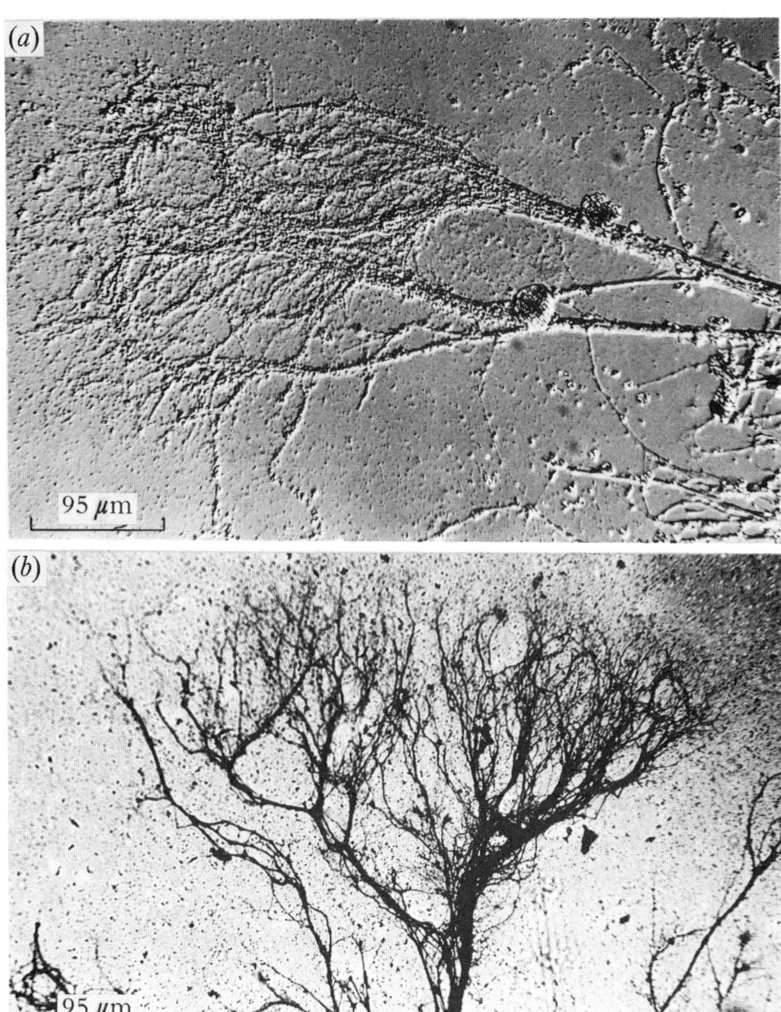

Plate 2 (*a*), (*b*). Large fibre bundles emerge from the protocerebrum (*a*) and from the 6th abdominal ganglion (*b*) of a 16-day embryo cultured *in vitro* for 20 days. In both instances the colossal fibre bundles dissociate at their distal ends in thin filaments. (*a*), living culture at the Nomarski microscope; (*b*), Cajal silver-stained preparation.

emerging as single filaments are seen in the migratory area around thoracic ganglia, while the large fibre bundles grow out from one or the other of the brain vesicles and from abdominal ganglia. Thus, as for the case of migrating nerve cells, it is tempting to extrapolate and suggest that individual nerve fibres are axons of motor neurons, whereas gregarious axons emerging as colossal bundles originate from the large and diversified cell populations of interneurons.

A third pattern of nerve fibre outgrowth which shares some properties in common with the two described above, but differs in other respects, is seen in the migratory area around embryonic explants of the frontal and ingluvial ganglia. Plate 3*b* and *c* shows the two ganglia after 15 days of culture. The fibres assemble in small bundles and are curled rather than straight like those shown in Plate 3*a*, which emerge from a thoracic ganglion. Intermingled with axons are glial cells.

Selectivity in nerve fibre outgrowth from thoracic and abdominal ganglia

When multiple explants of brain, ganglia, foregut segments are combined in the same culture vessel at a short distance from each other, they become interconnected by fibres branching out from brain and ganglia and also from sensory and visceral nerve cells which are part of the foregut. These findings raise the question whether the fibres, given multiple choices, would make preferential connections with some explants, or whether the connections between combined explants would occur at random. The results reported in another article (Levi-Montalcini & Chen, 1971) and summarised here, clearly favour the first alternative.

Thoracic and abdominal ganglia from 16-day embryos were combined in numbers of 10 to 12 in the same culture vessels. In other experiments, the same ganglia were combined with leg primordia and with segments of the alimentary tract from the same donors. Care was taken to position the explants at the same distance from each other – 0·8 or 1 mm. The cultures, many hundreds in number, were examined daily and again after discontinuation and staining with the Cajal silver technique. Fibres from adjacent ganglia show no preferential orientation toward each other until the end of the second week of culture, as shown in

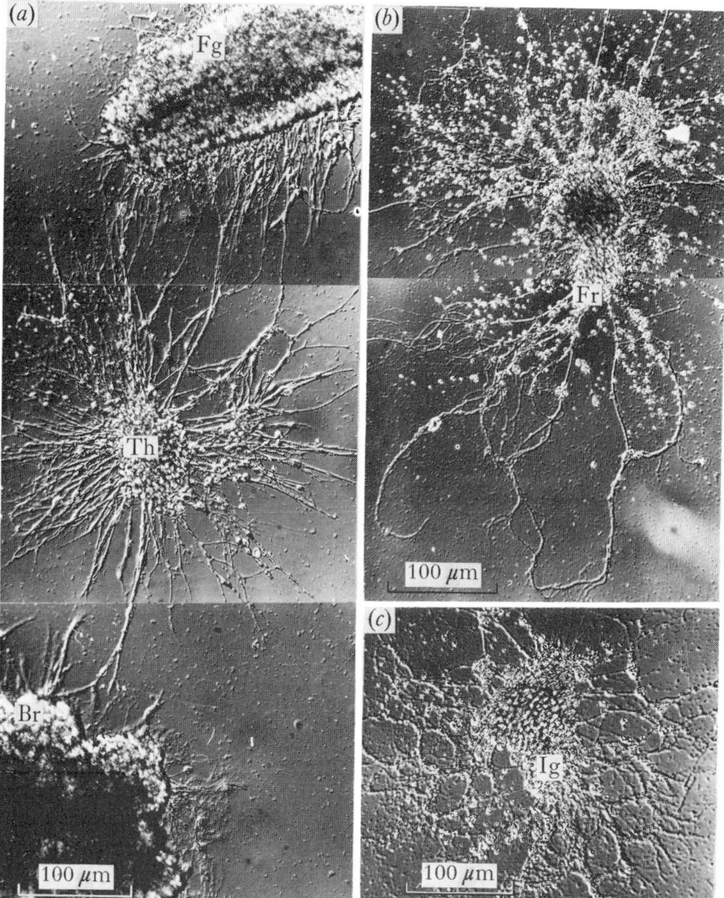

Plate 3. Photomicrographs (Nomarski) illustrating the pattern of nerve fibre outgrowth from a thoracic ganglion (Th) (*a*), a frontal ganglion (Fr) (*b*) and an ingluvial ganglion (Ig) (*c*) of 16-day embryos cultured *in vitro* for 15 days and photographed in living culture. Note radial outgrowth of fibres from thoracic ganglion making random connections with brain (Br) and foregut (Fg) from same donor. Large numbers of nerve cells are intermingled among axons. Fibres emerging from frontal and ingluvial ganglia are curled and exhibit a circular rather than radial pattern of growth. Intermingled with fibres are glial cells.

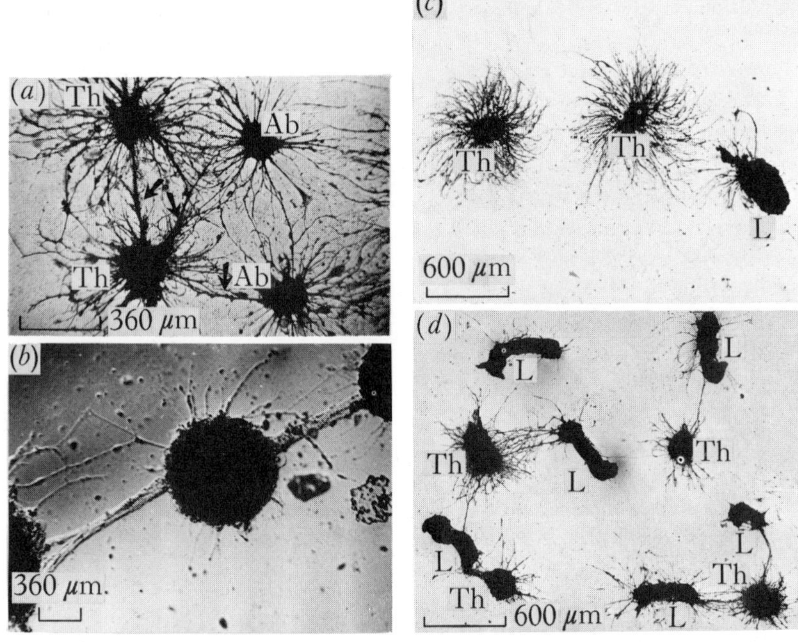

Plate 4. (*a*) Two thoracic (Th) and two abdominal ganglia (Ab) from a 16-day embryo cultured *in vitro* for 2 weeks, fixed and stained with Cajal silver technique. Arrows point to fibre bundles interconnecting facing explants. Note radial outgrowth of fibres from the ganglion contours.

(*b*) Photomicrograph (Nomarski) of living culture of 3 thoracic ganglia from a 16-day embryo cultured *in vitro* for 6 weeks. Large and compact fibre bundles interconnect facing ganglia, while nerve fibres growing out from other sectors of the same ganglia have undergone reabsorption.

(*c*) (*d*) Comparison of nerve fibre outgrowth from thoracic ganglia and leg primordia from 16-day embryos cultured *in vitro* for 2 weeks (*c*) still show no preferential orientation toward each other and leg primordium (L). In the 50-day old culture, all leg explants (L) are innervated by multiple fibre bundles emerging from thoracic (Th) ganglia. Both cultures stained with Cajal technique. ((*d*) from Levi-Montalcini & Chen, 1971.)

Plate 4*c*. At the beginning of the third week, fibres growing out from facing explants assemble in small bundles and direct their course toward matching ganglia, as shown in Plate 4*a*. In older cultures, such as that shown in Plate 4*b*, which was discontinued after 6 weeks *in vitro*, fibre bundles interconnecting pairing ganglia are of a remarkably large size and mimic to a surprising degree the connectives which interlink the same ganglia in the living organism. The similarity is further stressed by the fact that fibres growing out from other sectors of the ganglia have undergone reabsorption and only those which succeeded in establishing connections with adjacent ganglia have further increased in length and thickness.

A different situation occurs when the same ganglia are combined *in vitro* with leg primordia. In this case, as illustrated in Plate 4*d*, thoracic ganglia establish connections with leg tissues, usually through multiple fibre bundles which, in turn, resemble nerves emerging from the same ganglia *in vivo*. When thoracic ganglia are combined *in vitro* with other ganglia and with leg primordia, fibres emerging from different sectors of the ganglia make connections with both ganglionic and leg explants. Abdominal ganglia, while invariably connecting with like and unlike ganglia, only very seldom innervate leg explants.

These results suggest a different origin for axons with ganglia or leg tissues: the former would originate from interneurons, the latter from motor neurons. Experiments now in progress by one of us (J. S. Chen) bring additional evidence in favour of this hypothesis. Thoracic ganglia were divided by a horizontal cut into an anterior and a posterior half. It is known that the majority of leg motor neurons reside in the anterior half of these ganglia and interneurons are probably concentrated in the posterior half (see Young, this volume). The two half ganglia were then combined *in vitro* with leg primordia. Fibres emerging from the anterior half of the ganglia established connections with leg primordia, whereas no nerve connections were made with leg explants by fibres emerging from the posterior half.

While these systems are still too complex to give meaningful information on the forces which direct nerve fibres toward matching nerve cells or other tissues, they show that the insect nervous system, at variance with the vertebrate nervous system, retains upon explantation *in vitro* its structural organisation, thus offering the possibility of exploring problems of structure and function at the supercellular levels.

Neuronal nets from dissociated neurons in glial-free cultures

The finding that immature neurons migrating out from thoracic ganglia and, to a less extent, from brain and abdominal ganglia, lie on the surface of large glial cells was taken as evidence for the cell-to-cell interaction which characterises the relationship between nerve and glial cells and for the trophic role displayed by the latter on behalf of the former; such a role has in fact been considered as essential for the survival of nerve cells. In order to submit to a closer inspection the relationship between the two partners, we resorted to dissociation by mechanical procedures of brain and ganglia from 16-day embryos. The technique and the results are given in a previous article (Chen & Levi-Montalcini, 1970*a*). Here it suffices to say that nerve, but not glial, cells are able to withstand the trauma of the mechanical disruption of the nerve tissues. A plausible explanation for this differential effect on the two cell populations is to be found in the fact that nerve cells, loosely arranged in the cortical ring of brain and ganglia, disengage themselves easily from the soft matrix of the embryonic tissues, also on account of their round shape, while glial cells, which consist of a much more tenuous material and are very large and thin, undergo total disruption. The disappointment of not being able to explore the nerve–glial cell relationship was largely compensated by the availability of a pure nerve cell population uncontaminated by glial or other cells. Upon their dispersion in the chemically defined medium used in all previous investigations, the nerve cells and debris of glial cells attached to the glass surface. When cultured alone, even if in very large numbers, nerve cells produced short axons in the first three weeks of culture, but in longer-term cultures they showed manifest signs of deterioration and were discontinued at the beginning of the second month *in vitro*. An entirely different situation obtains when the same cells are cultured in combination with foregut or other embryonic tissues. Since the combined cultures of dissociated nerve cells and foregut segments were more extensively investigated than other combined cultures, the following description is based on the study of these nerve cell–foregut combinations.

A diagrammatic representation of one of these cultures kept *in vitro* for four months is shown in Fig. 1. The explant is sur-

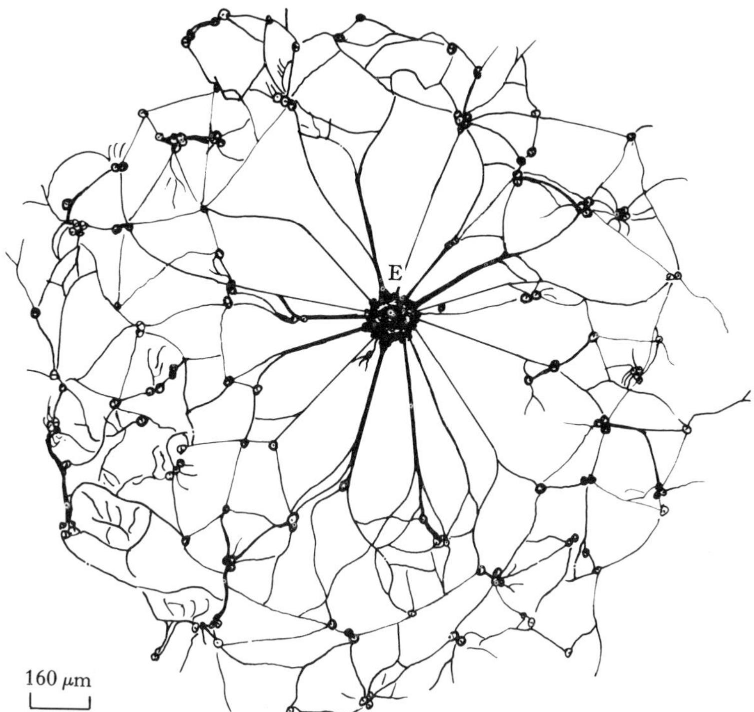

Fig. 1. Drawing of a 4-month old culture of dissociated nerve cells from a 16-day embryo around a foregut explant (E). (From Chen & Levi-Montalcini, 1970*a*.)

rounded by a fibrillar network produced by the dissociated nerve cells, which are seen in the nodal points of this net. From this net, organised in a concentric ring around the explants, nerve fibres assemble in bundles and take a straight, radially-oriented direction toward the explant, which is thus invaded by multiple fibre strands entering it throughout its periphery. Ultrastructural studies gave definite evidence for the total absence of glial cells among neurons and around axons in the fibrillar network and in the fibre bundles which enter into the explants. These electron microscopic studies also showed the normal appearance of nerve fibres in spite of the lack of glial mesaxons; neurotubules and neurofilaments are abundant at the axon hillock. Axon-to-axon synapses as well as axon-to-muscle cell junctions inside the explants exhibit the well known synaptic apparatus, consisting of synaptic vesicles and thickening of the presynaptic membranes. In a few instances it was also possible to see electron dense bodies in some fibres, suggestive of the presence of neurosecretory material (Chen & Levi-Montalcini, 1970*b*). The possibility, however, that these glial-free neurons may be defective in some enzyme systems was recently raised by Schlapfer *et al.* (1972). These authors explored the cholinesterase and cholin acetyltransferase activities in whole brain explants and in dissociated nerve cells cultured according to our 'two floor system', which consists in dispersing nerve cells on a different coverslip, but in the same culture vessel together with foregut segments. The activity of these two enzymes, assayed in whole brain and ganglion explants, increased considerably after two weeks of culture, while no significant rise was detected in dissociated nerve cells after the same length of time. The authors concluded that neuronal differentiation in the dissociated nerve cells, although morphologically impressive, is enzymologically incomplete (p. 543).

Such a conclusion would have been much more justified if enzymatic activity had been measured in 3- to 4-month old cultures rather than after such a short time of incubation. Our extensive experience with dissociated nerve cells showed in fact that axonal growth is very slow during the first month of culture and then progresses at a much faster rate in subsequent months. Cultures examined between the second and third weeks of culture are in fact far from being impressive, even from a morphological viewpoint, and would convey the impression that these cells failed to differentiate under the conditions of culture. It is only at later stages that they acquire morphological features suggestive of a normal differentiation of the cell perikaryon as well as of its axon. Pending more extensive experiments and enzyme measurements in long-term cultures, the question is at present unsettled.

A few words should also be said about the maintenance and differentiating effects elicited by foregut explants on dissociated nerve cells. Recently we explored the effects of explants of brain and ganglia combined *in vitro* with dissociated nerve cells. Both tissues, but to a major extent the ganglia, enhance axonal growth and survival of dissociated nerve cells (Levi-Montalcini & Aloe, 1972). These observations, already mentioned in the first part of this article, would suggest a conditioning effect of the culture medium by actively growing embryonic tissues. Release of humoral conditioning factors by growing cells could also account for the finding that foregut segments cultured alone undergo rapid and progressive deterioration, whereas the same segments undergo differentiation and become contractile when cultured in combination with dissociated nerve cells (Chen & Levi-Montalcini, 1970*b*). In this case, however, it is also possible that such a trophic effect results from direct contact of the entering nerve fibres with muscle cells, rather than from an aspecific conditioning effect on the medium. Such an *in-vitro* effect in vertebrate cultures has in fact been described (Crain & Peterson, 1967; Crain, Peterson & Bornstein, 1968).

The insect neurovisceral system

The object of this study was to explore the relationship between the ingluvial ganglion, which innervates the anterior segments of the alimentary tract and controls its motility and the muscle cells innervated by this ganglion. The results reported in a recent article (Aloe & Levi-Montalcini, 1972*a*) are briefly summarised here.

Multiple foregut segments from 16-day embryos, with the small ovoidal ingluvial ganglion attached, were cultured *in vitro* in close proximity to each other. In some instances, dissociated nerve cells were also added to improve the culture medium. In the second month of culture, nerve fibres emerging from the ingluvial ganglia branched out from the explants and directed their course toward the same ganglion of matching explants. Muscle cells from the circular and longitudinal layer of the foregut, in turn, moved out from the explants and lined on the surface of the axons of ingluvial nerve cells, or scattered around the explants, intermingled with nerve fibres. These thin, spindle-shaped cells were identified as myocytes on account of their spontaneous contractility and electron microscopic studies which showed the presence of myosin and actin filaments (Aloe & Levi-Montalcini, 1972*a*). Time-lapse cinematography and observations of living cultures with the Nomarksi microscope revealed new features of the muscle–nerve fibre relationship; the migrating myocytes show in fact a marked affinity for axons of ingluvial neurons. They adhere to these fibres

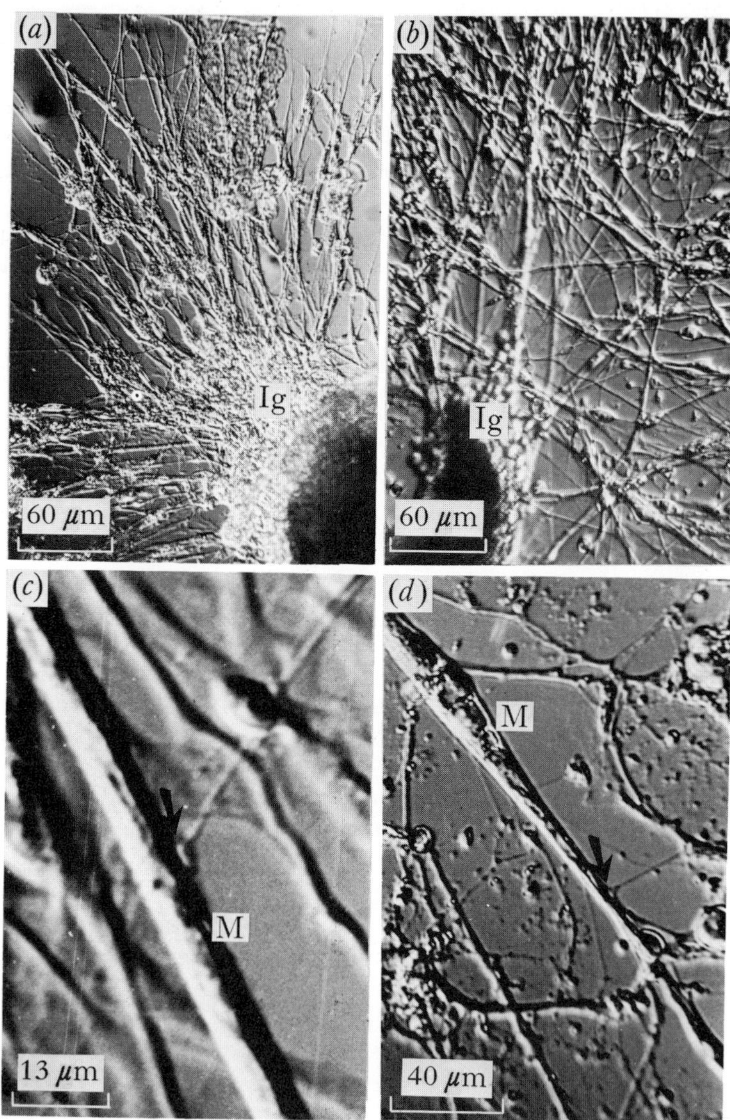

Plate 5. Photomicrographs (Nomarski) of a 3-month old living culture of oesophagus and attached ingluvial ganglion from 16-day embryos cultured in presence of other like explants and of dissociated nerve cells. Ig, ingluvial ganglion; M, myocytes. Arrows in (*c*) (*d*) point to junction sites of nerve fibres with myocytes. Additional explanation in text.

or attach to their distal tips. A material contact between muscle cells and nerve fibres is suggested by the observation that the latter are set in motion by contractile myocytes, as if pushed and pulled by these cells. As a result, the nerve fibres seem to contract and relax like myocytes, while obviously this is a passive rather than active type of motility. Plate 5*a–d* shows one of these 3-month old cultures at low and high magnification.

These findings bring to light an aspect of muscle–nerve fibre relationship which is not noticeable in the living organism, namely the strong affinity and thigmotropism of muscle cells for the visceral nerve fibres which provide their innervation. Since in the living organism the muscle cells of organs such as the alimentary tract are anchored in place in the dense texture of the viscera, this property cannot manifest itself, while it becomes apparent when the cells regain their freedom under the conditions provided by culture in liquid media. These cultures are now used by one of us (L. Aloe) to explore the effects of pharmacological agents and putative neurotransmitters on this dual muscle–nerve fibre system under conditions which seem to be far more amenable to exploration than those of the living organism.

The nymphal and adult insect nervous system

The results obtained with the embryonic nervous system encouraged us to extend the analysis to the nymphal and adult insect nervous system. The studies centred on the stomatogastric and neuroendocrine organ complexes. In both instances the fully differentiated ganglia and organs adapted remarkably well to the conditions of culture, which were the same as those devised for embryonic tissues. In the following, we will briefly consider some of the results obtained with both systems and the perspectives opened by these investigations.

The frontal ganglion

This large, triangular-shaped ganglion, which represents one of the main components of the insect stomatogastric system, is connected with the tritocerebrum through two robust nerves known as the frontal connectives, and with the ingluvial ganglion through the oesophageal nerve. The ganglion and the above two nerves are

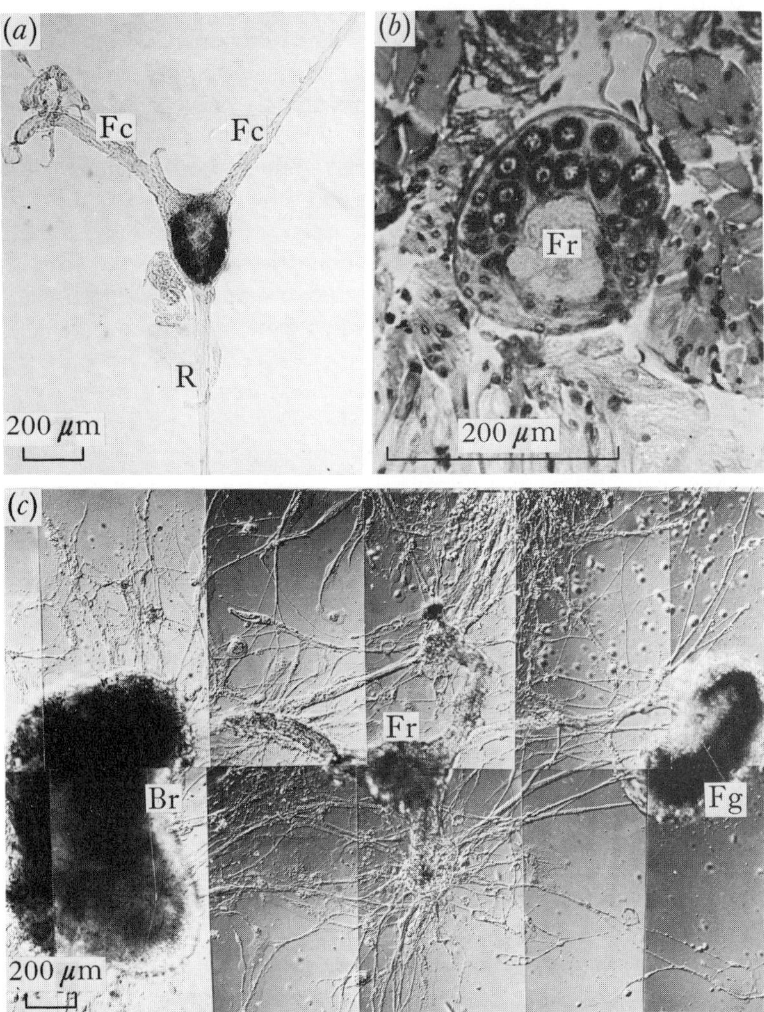

Plate 6. (*a*) Whole mount of a frontal ganglion dissected out from a 7th instar nymph. Fc, frontal connectives; R, recurrent nerve.

(*b*) Transverse section of frontal ganglion (Fr) stained with haematoxylin and eosin, from 7th instar nymph.

(*c*) Photomicrograph (Nomarski) of living culture of frontal ganglion from 7th instar nymph cultured *in vitro* for 2 months in presence of a whole brain (Br) explant and foregut segment (Fg) from a 16-day embryo. Note the multiple fibre bundles emerging from the ganglion and establishing connections with adjacent explants.

shown in Plate 6*a*. According to Bullock & Horridge (1965), the ganglion assembles motor and sensory functions, being on the sensory pathway from the alimentary tract to the brain and at the same time possessing distinct features of a motor centre interposed between the brain and the ingluvial and proventricular ganglia which control the motility of the anterior alimentary tract. Besides its motor and sensory functions, the ganglion exerts a control on metabolic functions through its multiple and complex connections with neurosecretory brain centres and with the adjacent neuro-endocrine complex; its ablation is, according to some authors, incompatible with life (Clarke & Gillot, 1965; Clarke & Langley, 1963*a*, *b*). The neurons, in number from 95 to 98, are aligned in two rows around a central neuropile, as shown in Plate 6*b*. From a structural and physiological viewpoint, the ganglion presents a most favourable and attractive object of investigation; it is in fact easily accessible to exploration and its large neurons located in the ganglion marginal ring are ideally placed for electrophysiological and pharmacological studies of individual units, as well as of the entire small and compact nerve cell population.

The observations to be reported below represent a first attempt to submit this ganglion to an *in-vitro* analysis for the purpose of gaining information on these cells at the structural and ultra-structural levels and, at the same time, providing a baseline for electrophysiological, biochemical and pharmacological studies. The results reported in detail in a previous article (Aloe & Levi-Montalcini, 1972*b*) are briefly summarised here.

The frontal ganglion of nymphal and adult insects survives in excellent condition and produces a profuse nerve fibre outgrowth when cultured *in vitro* in combination with dissociated nerve cells or embryonic organs from 16- to 18-day donors. It undergoes instead progressive deterioration when one or even multiple explants of the ganglion are cultured in absence of embryonic nerve cells or tissues. It is of particular interest to note that the production of nerve fibres by the fully differentiated neurons is of the same range and perhaps even more profuse than that of embryonic nerve tissue. While axons branching out from the embryonic frontal ganglion assemble in small bundles which run in a circular fashion around the explant (Plate 3*c*), they assemble in large bundles in the present cultures and establish connection with adjacent embryonic organs or axons of dissociated nerve cells.

A typical 2-month old culture is shown in Plate 6*c*. The ganglion, of considerable size, connects with two adjacent brain and foregut embryonic explants.

The neuro-endocrine complex

This organ complex which plays a key role in morphogenetic events at different stages in the insect's life is also endowed with properties similar in many respects to those of the vertebrate neurohypophyseal system. In both vertebrates and invertebrates, this organ complex consists of a neural and non-neural component. In the case of insects, the paired corpora cardiaca have a neural origin and share some features in common with the neurohypophysis, although at a closer inspection marked differences in the structural organisation of the two glands are apparent (Bern, 1971; Bern & Hagadorn, 1965; Scharrer & Scharrer, 1963; Smith, 1968). Structural, ultrastructural and biochemical studies have centred on the analysis of the main cell constituents of corpora cardiaca and of the products released by these glands and identified as polypeptides endowed with manifold and diversified biological activities (Brown, 1965; Highnam & Goldsworthy, 1972; Mordue & Goldsworthy, 1969; Natalizi & Frontali, 1966; Steele, 1961). The corpora allata, which consist of two small, round-shaped glands, possess most important endocrine functions which were also the object of extensive investigation in recent years (Bell, 1969; Bowers, 1971; Fain-Maurel & Cassier, 1969; King, Aggarwal & Bodenstein, 1966; Lea & Thomsen, 1969; Odhiambo, 1966*a*, *b*; Scharrer, 1964, 1970; Staal, 1971; Thomsen & Thomsen, 1969, 1970). Each corpus allatum receives axons from the neurosecretory cells located in the protocerebrum; the axons, filled with neurosecretory material, pass through the corpora cardiaca before ending in the corpora allata. To what extent the products synthesised in the protocerebral neurosecretory cells are modified in the corpora cardiaca, is at present not known.

These *in-vitro* studies which are at present actively pursued by one of us (K. R. Seshan) centred mainly on the analysis of corpora cardiaca and of their nerve cell population and fibre constituents in long-term cultures. Neurons of these glands differ from typical neurons in possessing stainable material, apparently synthesised in the cell perikarya and passed along the short and bulbous axons. Nerve fibres differ from true axons in the fact that they contain ribosomes (Smith, 1968; Smith & Smith, 1966). It is generally assumed that these axon-like processes do not extend beyond the confines of the cardiacum glands. Besides these cells, which represent the largest cell population of the corpora cardiaca, there are interstitial cells which have many characteristics in common with brain and ganglion glial cells. According to some authors, a small population of true neurons is also present (Bern & Hagadorn, 1965; Gabe, 1966; Meola & Lea, 1972; Scharrer, 1962).

The *in-vitro* system seemed to offer an ideal condition to explore the structural and functional properties of the neuroglandular cells when separated from the protocerebral neurosecretory cells, which are credited with exerting a control on their function. The present studies, reported in detail in another publication (Seshan & Levi-Montalcini, 1971), revealed some new features of these cells which stress their similarity with conventional neurons. Upon their explantation *in vitro*, a large number of fibres, identified at the light and electron microscope as axons, branch out into the medium as isolated fibres, or assembled in fibre bundles which bridge the distance between these glands and adjacent organs. Connections were seen between these fibres and embryonic explants of brain, ganglia, foregut and heart. Electron microscopic studies (unpublished experiments by K. R. Seshan) showed that axons emerging from neuroglandular cells still contain a sparse number of electron dense bodies after three months of culture. The problem of whether these fibres are able to make conventional synapses with tissues of embryonic organs, or whether they end blindly and release neurosecretory material in the same fashion as neurosecretory fibres in the living organism, is now under investigation. Likewise the electrical properties of these axons is now under investigation in our laboratory by Dr R. Provine.

Concluding remarks

The aim of this article, as stated in the introduction, was to present a panoramic view rather than to elaborate in detail on results of investigations which should be considered as preliminary excursions in a field which opens for the first time to exploration. While the tissue culture techniques are now routinely used in many laboratories to grow insect cells and also to study the effects of hormones and other humoral agents on target cells and intact organs, practically no efforts were made to use the same techniques to gain information on the insect nervous system. An exception is represented by the investigations performed by Marks & Reinecke who, since 1965, have reported on the successful growth *in vitro* of leg regenerates from nymphs of the cockroach *Leucophaea maderae*, cultured in combination with the prothoracic ganglion, prothoracic gland and also, in subsequent experiments, with brain tissues from the same donors (Marks, 1968, 1970; Marks & Reinecke, 1965; Marks, Reinecke & Leopold, 1968). In these articles they described the action of agents released by nerve tissues and glands on the regeneration of leg tissues. As a sideline of these investigations, they touched on the nerve fibre outgrowth from nerves emerging from the cut stump of the leg and also on nerve fibre outgrowth from the thoracic ganglion.

The results reported in this article show that the insect nervous system lends itself remarkably well to an *in-vitro* analysis from its early differentiation to maturity. It is, in fact, much more amenable to such analysis than the vertebrate nervous system, perhaps in view of its much smaller size and organisation in compartments which are, to a large extent, independent of each other at the structural and functional levels, as well as its easily met nutritional requirements.

The remarkable ability of insect nerve cells and even of intact brain and ganglia to survive *in vitro* for many months in a medium which has very little in common with that provided by the living organism, has exceeded our expectations to such a point as to become a matter of wonder and perplexity. Both embryonic and adult nerve tissues not only survived for many months in a chemically defined medium, but also produced vigorous nerve fibre outgrowth and established connections with adjacent embryonic organs or dissociated nerve cells. We are now restricting, rather than extending, the area of investigation to avoid the danger of spreading too thin in this field, which lures us like all unexplored fields, irrespective of what they hold in store for the explorers. To this aim, we plan to shift the approach from the structural to the ultrastructural, electrophysiological and biochemical levels. Preliminary excursions have already been made and the results are encouraging. Thus, four years after we first approached the insect nervous system with rather vague projects in mind and only a rudimentary knowledge of its main structures, we feel that an optimistic outlook on the possible development of these investigations is well justified. The insect nervous system seems in fact to provide a convenient model for studying problems of nerve cell structure and function, as well as of its organisation at the supercellular level.

Ending in the same light and cheerful vein as did Benzer after presenting the results of his remarkable studies on the behaviour of fruit flies explored with genetic tools (Benzer, 1971), we can conclude, as he did, that in any case it is fun.

This work was supported in part by grants from the US Public Health Service (NS-03777) and from the National Science Foundation (GB-16330X).

References

Abercrombie, M. (1958). Exchanges between cells. In *A Symposium on the Chemical Basis of Development*, ed. W. D. McElroy & B. Glass, pp. 318–28. Baltimore: Johns Hopkins Press.

Aloe, L. & Levi-Montalcini, R. (1972*a*). Interrelation and dynamic activity of visceral muscle and nerve cells from insect embryos in long-term cultures. *Journal of Neurobiology*, **3**, 3–23.

(1972*b*). *In-vitro* analysis of the frontal and ingluvial ganglia from nymphal specimens of the cockroach *Periplaneta americana*. *Brain Research*, **44**, 147–63.

Bell, W. J. (1969). Dual role of juvenile hormone in the control of yolk formation in *Periplaneta americana*. *Journal of Insect Physiology*, **15**, 1279–90.

Benzer, S. (1971). From gene to behaviour. *Journal of the American Medical Association*, **218**, 1015–22.

Bern, H. A. (1971). The status of neuroendocrine structures in invertebrates analogous with those in vertebrates. In *Memoirs of the Society for Endocrinology*, no. 19, Subcellular Organization and Function in Endocrine Tissues, 843–51. London: Cambridge University Press.

Bern, H. A. & Hagadorn, I. R. (1965). Neurosecretion. In *Structure and Function in the Nervous Systems of Invertebrates*, ed. T. H. Bullock & G. A. Horridge, pp. 353–429. San Francisco & London: Freeman.

Bowers, W. S. (1971). Insect hormones and their derivatives as insecticides. *Bulletin of the World Health Organization*, **44**, 381–9.

Brown, B. E. (1965). Pharmacologically active constituents of the cockroach corpus cardiacum: resolution and some characteristics. *General and Comparative Endocrinology*, **5**, 387–401.

Bullock, T. H. & Horridge, G. A. (1965). *Structure and Function in the Nervous Systems of Invertebrates*, vol. II, pp. 802–1592. San Francisco & London: Freeman.

Chen, J. S. & Levi-Montalcini, R. (1969). Axonal outgrowth and cell migration *in vitro* from nervous system of cockroach embryos. *Science*, **166**, 631–2.

(1970*a*). Axonal growth from insect neurons in glia-free cultures. *Proceedings of the National Academy of Sciences, USA*, **66**, 32–9.

(1970*b*). Long-term cultures of dissociated nerve cells from the embryonic nervous system of the cockroach *Periplaneta americana. Archives italiennes de Biologie*, **108**, 503–37.

Clarke, K. U. & Gillot, C. (1965). Relationship between the removal of the frontal ganglion and protein starvation in *Locusta migratoria*, L. *Nature, London*, **208**, 808–9.

Clarke, K. U. & Langley, P. A. (1963*a*). Studies on the initiation of growth and moulting in *Locusta migratoria migratoriodes*, R & F. II. The role of the stomatogastric nervous system. *Journal of Insect Physiology*, **9**, 363–73.

(1963*b*). Studies on the initiation of growth and moulting in *Locusta migratoria migratoriodes*, R & F. III. The role of the frontal ganglion. *Journal of Insect Physiology*, **9**, 411–21.

Crain, S. M. & Peterson, E. R. (1967). Onset and development of functional interneuronal connections in explants of rat spinal cord ganglia during maturation in culture. *Brain Research*, **6**, 750–62.

Crain, S. M., Peterson, E. R. & Bornstein, M. B. (1968). Formation of functional interneuronal connections between explants of various mammalian central nervous tissues during development *in vitro*. In *Ciba Foundation Symposium on Growth of the Nervous System*, ed. G. E. W. Wolstenholme & M. O'Connor, pp. 13–31. London: Churchill.

Edwards, J. S. (1969). Postembryonic development and regeneration of the insect nervous system. *Advances in Insect Physiology*, **6**, 97–137.

Fain-Maurel, M. A. & Cassier, P. (1969). Étude infrastructurale des corpora allata de *Locusta migratoria migratoriodes* (R. et F.), phase solitaire, au cours de la maturation sexuelle et des cycles ovariens. *Comptes Rendus Hebdomadaires des Séances de l'Académie des Sciences, Paris*, Ser. D., **268**, 2721–3.

Florey, E. (1967). Introductory remarks. In *Invertebrate Nervous Systems*, ed. C. A. G. Wiersma. Chicago: University of Chicago Press.

Gabe, M. (1966). *Neurosecretion*. Oxford, London & New York: Pergamon Press.

Grace, T. D. C. (1962). Establishment of four strains of cells from insect tissues grown *in vitro*. *Nature, London*, **195**, 788–9.

(1966). Establishment of a line of mosquito (*Aedes aegypti* L.) cells grown *in vitro*. *Nature, London*, **211**, 366–7.

Highnam, K. C. & Goldsworthy, G. J. (1972). Regenerated corpora cardiaca and hyperglycemic factor in *Locusta migratoria*. *General and Comparative Endocrinology*, **18**, 83–8.

King, R. C., Aggarwal, S. K. & Bodenstein, D. (1966). The comparative submicroscopic cytology of the corpus allatum–corpus cardiacum complex of wild type and *fes* adult female *Drosophila melanogaster*. *Journal of Experimental Zoology*, **161**, 151–76.

Landureau, J. C. (1966). Cultures *in vitro* de cellules embryonnaires de Blattes (insectes dictyoptères). *Experimental Cell Research*, **41**, 545–56.

(1969). Étude des exigences d'une lignée de cellules d'insectes (souche EPa). II. Vitamines hydrosolubles. *Experimental Cell Research*, **54**, 399–402.

Landureau, J. C. & Jollès, P. (1969). Étude des exigences d'une lignée de cellules d'insectes (souche EPa). I. Acides amines. *Experimental Cell Research*, **54**, 391–8.

Lea, A. O. & Thomsen, E. (1969). Size-independent secretion by the corpus allatum of *Calliphora erythrocephala. Journal of Insect Physiology*, **15**, 477–82.

Levi-Montalcini, R. (1963). Growth and differentiation in the nervous system. In *The Nature of Biological Diversity*, ed. J. M. Allen, pp. 261–95. New York: McGraw-Hill.

Levi-Montalcini, R. & Aloe, L. (1972). Neuronal nets and nerve cell interactions *in vitro* in insect systems. *In Vitro*, **8**, 178–91.

Levi-Montalcini, R. & Chen, J. S. (1969). *In vitro* studies of the insect embryonic nervous system. In *Cellular Dynamics of the Neuron*, ed. S. H. Barondes, pp. 277–98. New York: Academic Press.

(1971). Selective outgrowth of nerve fibers *in vitro* from embryonic ganglia of *Periplaneta americana. Archives italiennes de Biologie*, **109**, 307–37.

Lumsden, C. E. (1968). Nervous tissue in culture. In *The Structure and Function of Nervous Tissue*, ed. G. H. Bourne, vol. 1, pp. 67–140. New York: Academic Press.

Marks, E. P. (1968). Regenerating tissues of the cockroach *Leucophaea maderae*: effects of humoral stimulation *in vitro. General and Comparative Endocrinology*, **11**, 31–42.

(1970). The action of hormones in insect cell and organ cultures. *General and Comparative Endocrinology*, **15**, 289–302.

Marks, E. P. & Reinecke, J. P. (1965). Regenerating tissues from the cockroach *Leucophaea maderae*: effects of endocrine glands *in vitro. General and Comparative Endocrinology*, **5**, 241–7.

Marks, E. P., Reinecke, J. P. & Caldwell, J. M. (1968). Cockroach tissue *in vitro*: a system for the study of insect cell biology. *In Vitro*, **3**, 85–92.

Marks, E. P., Reinecke, J. P. & Leopold, R. A. (1968). Regenerating tissues from the cockroach *Leucophaea maderae*: nerve regeneration *in vitro. Biological Bulletin. Marine Biology Laboratory, Woods Hole*, **135**, 520–9.

Meola, S. M. & Lea, A. O. (1972). The ultrastructure of the corpus cardiacum of *Aedes sollicitans* and the histology of the cerebral neurosecretory system of mosquitoes. *General and Comparative Endocrinology*, **18**, 210–34.

Mordue, W. & Goldsworthy, G. J. (1969). The physiological effects of corpus cardiacum extracts in locusts. *General and Comparative Endocrinology*, **12**, 360–9.

Natalizi, G. M. & Frontali, N. (1966). Purification of insect hyperglycaemic and heart accelerating hormones. *Journal of Insect Physiology*, **12**, 1279–87.

Odhiambo, T. R. (1966*a*). The fine structure of the corpus allatum of the sexually mature male of the desert locust. *Journal of Insect Physiology*, **12**, 819–28.

(1966*b*). Ultrastructure of the development of the corpus allatum in the adult male of the desert locust. *Journal of Insect Physiology*, **12**, 995–1002

Scharrer, B. (1962). Neurosecretion. The fine structure of the neurosecretory system of the insect Leucophaea maderae. *Memoirs of the Society for Endocrinology*, **12**, 89–97.

(1964). Histophysiological studies on the corpus allatum of *Leucophaea maderae.* IV. Ultrastructure during normal activity cycle. *Zeitschrift für Zellforschung und Mikroskopische Anatomie*, **62**, 125–48.

(1970). Ultrastructural study of the sites of origin and release of a cellular product in the corpus allatum of insects. *Proceedings of the National Academy of Sciences, USA*, **66**, 244–5.

Scharrer, E. & Scharrer, B. (1963). *Neuroendocrinology*. New York & London: Columbia University Press.

Schlapfer, W. T., Haywood, P. & Barondes, S. H. (1972). Cholinesterase and choline acetyltransferase activities develop in whole explant but not in dissociated cell cultures of cockroach brain. *Brain Research*, **39**, 540–4.

Schneider, I. (1964). Differentiation of larval *Drosophila* eye-antenna discs *in vitro*. *Journal of Experimental Zoology*, **156**, 91–104.

Seshan, K. R. & Levi-Montalcini, R. (1971). *In vitro* analysis of corpora cardiaca and corpora allata from nymphal and adult specimens of *Periplaneta americana*. *Archives italiennes de Biologie*, **109**, 81–109.

Smith, D. S. (1968). *Insect Cells: their Structure and Function*. Edinburgh: Oliver & Boyd.

Smith, U. & Smith, D. S. (1966). Observations on the secretory processes in the corpus cardiacum of the stick insect, *Carausius morosus*. *Journal of Cell Science*, **1**, 59–66.

Staal, G. B. (1971). Practical aspects of insect control by juvenile hormone. *Bulletin of the World Health Organization*, **44**, 391–4.

Steele, J. E. (1961). Occurrence of a hyperglycaemic factor in the corpus cardiacum of an insect. *Nature, London*, **192**, 680–1.

Thomsen, E. & Thomsen, M. (1969). Fine structure of the corpus allatum of the female *Calliphora*, with special reference to hormone formation. *General and Comparative Endocrinology*, **13**, 534. (Abstract).

(1970). Fine structure of the corpus allatum of the female blow-fly *Calliphora erythrocephala*. *Zeitschrift für Zellforschung und Mikroskopische Anatomie*, **110**, 40–60.

Gradients and the developing nervous system

C. M. Bate and P. A. Lawrence

Introduction

Gradients have been widely canvassed as the source of the information whereby developing cells become polarised and assess their position within a morphogenetic field (Child, 1941; Wolpert, 1971). Theories of this kind have been applied to the development of hydroids, sea urchins and planarians. It has been suggested that gradients may regulate the formation of patterned connections within the developing nervous system (e.g. Jacobson, 1970). The most complete model is of an antero-posterior gradient within the insect segment; it accounts for the oriented deposition of cuticular structures and the regional differentiation of the integument (Lawrence, 1970). We shall examine this hypothesis as a model for the generation of polarity and positional information within the nervous system and as the basis of a mechanism which regulates the central connections of the sensilla in the insect epidermis.

Despite the complexity of the nervous system, the morphogenetic problem which it presents is of a conventional kind. In a recent discussion, Wolpert (1971) has emphasised that it should be a 'maxim to avoid inferring developmental mechanisms from final form or pattern'. In this view the instructions to build a system such as the neuropile of an insect ganglion are easier to describe than the final structure itself. Indeed, within the insect nervous system, evidence of the fine anatomy of motor neurons (Burrows, 1973) suggests that there may be many different routes to similar physiological solutions. Developing nerve cells therefore, do not have access to a blueprint of their final structure; their oriented growth and the establishment of particular connections depends on an interaction between genetically coded instructions and information in their environment. In particular, this information must include *polarity*, to direct the growing axon and an *addressing system* whereby an appropriate choice of terminals may be made.

The gradient in the insect segment

Both kinds of information are, in principle, present in the epidermal gradient of the insect segment. The experimental manipulation of the ripple pattern in the adult cuticle of *Rhodnius* (Locke, 1959) has shown that there are differences between insect epidermal cells which are graded in the antero-posterior axis. The orientation of the cuticular ripples indicates the polarity of the cells which secrete them. Locke rotated or transplanted squares of the larval integument and examined the polarity of the cells in the adult. He found that although the polarity of the rotated piece was partly maintained there was some interaction with the host tissue at the margin of the graft, and this interaction was expressed in the deposition of ripples with an altered orientation. He also showed that when normally oriented pieces of integument were transplanted to a more anterior or more posterior region of the segment the pattern of ripples was disturbed, and the further the graft was displaced the more disturbed the adult pattern became. Because transplantation in the medio-lateral axis or to equivalent sites in other segments had no effect on the ripple orientation, he deduced that there was an antero-posterior gradient which was repeated in each segment and that when cells were displaced within this gradient their subsequent orientation depended on the size and direction of the differences between them and surrounding cells. When cells from different values of the gradient were apposed the cellular polarity of both graft and host were altered by this interaction.

Although the existence of a gradient is deduced from its effects on cell polarity, it is also a potential source of information about position in the relevant axis. *The word gradient is often misapplied to developmental phenomena where only polarity has been observed. A gradient implies a scalar field, and strictly the word should only be used where a scalar, such as the value of concentration at a point, and a vector, such as the local direction of the maximal slope of concentration have been experimentally demonstrated* (Crick, 1971). Later experiments have shown that this information is used by the epidermal cells (Marcus, 1962; Stumpf, 1968; Lawrence, 1972). For example, Stumpf, who studied the regional characteristics of the pupal cuticle in *Galleria mellonella*, showed that the patterns of pupal integument produced by rotated or transplanted fragments of larval epidermis coincided precisely with the pattern of contours predicted on her theory of an antero-posterior gradient. She concluded that the response of competent cells was dependent on the local concentration of a diffusible substance which was produced at one segment margin.

In an attempt to discover more about the nature of the gradient, we have repeated and extended Locke's study and compared our results with the results of model experiments performed on a computer (Lawrence, Crick & Munro, 1972). Our main conclusions from this study are:

(i) There is a monotonic gradient of positional information, parallel with the antero-posterior axis which is repeated in each unit of developmental organisation (a segment) and is maintained between the boundaries of that unit.

(ii) The polarity of the epidermal cell, as expressed in the orientation of anisotropic structures in the cuticle, is dependent on the local direction of steepest slope of the gradient of positional information.

(iii) The gradient behaves as a concentration gradient of a diffusible substance (a morphogen) such that when cells from a high part of the gradient are placed next to those from a lower part diffusion occurs between them and the gradient value becomes intermediate. Usually this process will cause changes in the local slope and therefore of the polarities of the cells.

(iv) Because the rotation experiments produce patterns of polarity which are stable over many weeks (equilibrium being lost only when cell divisions occur), we propose that each cell is 'set' to a certain value in the concentration gradient and attempts to maintain that concentration even when it is moved. This set value is read directly from the concentration of the morphogen in the cell, at a particular point in the cell cycle.

Thus information as to polarity and position is generated in such a gradient and, since the daughter cells accept their set level from their surroundings and not from the parent cell alone, growth is accommodated within it. Is there evidence that information is produced in a similar way in the nervous system?

Gradients in the vertebrate visual system

The development of the insect model depended on the use and interpretation of polarised structures in the cuticle. In the projection of the vertebrate visual system from the retina to the brain there is an analogous expression of polarity. Each point in the visual field is connected to an equivalent point on the optic tectum by the axons of retinal ganglion cells. In this way the axes of the retina are projected onto equivalent axes in the tectum and polarities in the retina and tectum can be deduced by mapping the projection of any two points. Sharma & Gaze (1971) have recently published the results of rotation experiments which are analogous to those previously carried out on the epidermis of the insect segment. Cutting connections between the graft and the eye, they rotated pieces of the tectum of adult goldfish through ninety degrees and allowed the visual projection to regenerate. After nine to eleven months they examined the projections in the operated fish (Fig. 1*c*). We might expect that these would show one of two results: either the positional information in the tectum would be completely stable (the tectal cells being completely determined) giving the projection shown in Fig. 1*a*, or the rotation might have no effect at all, the positional information in the rotated patch and hence the subsequent connections between it and regenerating retinal axons being subject only to the orientation of the tectum as a whole. In the three fish which gave clear patterns the projections conformed to neither of these predictions (Fig. 1*c*). The polarities had become modified in an orderly manner, so that the patterns of connections could be compared with those which are produced by ninety degree rotations of fragments of insect epidermis (Fig. 1*b*). While there are a number of possible explanations (such as rotation back of the entire graft) these experiments are suggestive. They raise the question whether the polarity of the insect epidermis and the visual projection of vertebrates might depend on similar processes.

In a discussion of the amphibian visual system in normal development and after experimental interference, Gaze & Keating (1972) concluded that 'interconnecting structures tend to match up as systems rather than as a series of discrete independent subunits'. In many experiments this is expressed by the growth of a limited retinal input to spread in an orderly manner over the whole tectal

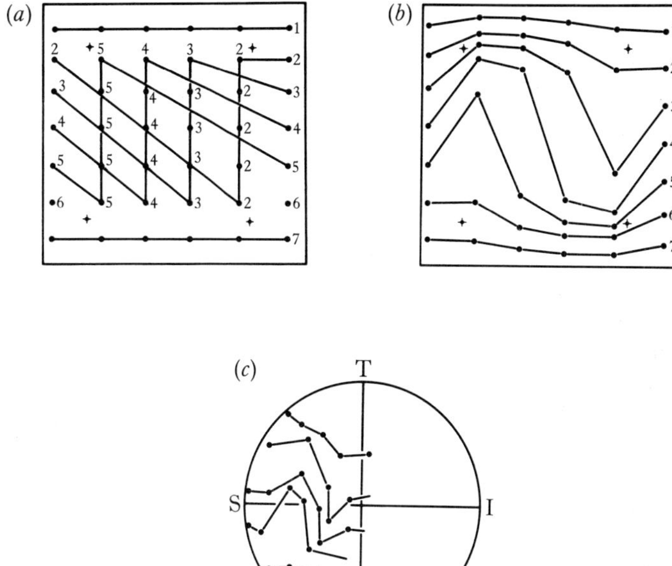

Fig. 1. Rotation of grafts through 90°; a comparison between the insect cuticle and the goldfish retinotectal projection.

(*a*) The expected pattern if the graft projection develops according to its prospective fate. Notice the crossing of lines. (*b*) Contour map at equilibrium after 90° rotation of a square of a concentration gradient ($K = 15$; Lawrence, Crick & Munro, 1972). The diffusion has resulted in a concentration landscape where one can draw straight lines between points at the same concentration without any lines crossing. (*c*) Retinotectal projection after rotation of a piece of goldfish tectum 90° clockwise (after Sharma & Gaze, 1971).

surface. Similarly, during the normal course of development (Gaze, Chung & Keating, 1972) the entire retina projects across the entire tectum at most stages, although the growth of the retina is topographically quite different from that of the tectum (Straznicky & Gaze, 1972). These observations imply that a given retinal ganglion cell projects to one part of the tectum at one stage of development and subsequently shifts to another part, although the order of the projection is maintained from the earliest stage. In other words, the affinity of retinal and tectal cells which causes particular points in the visual field to be represented at predictable points in the tectum of the adult changes, whereas the order within the retina and the tectum remains the same.

It is tempting to draw a parallel here with the growth of the insect segment. In the insect epidermis, the cells have a 'memory' – their set value within the segmental gradient of concentration. During proliferation of the epidermis mitoses occur all over the segment so that the set level and the actual level remain similar to each other as growth continues. If, as in the retina, growth were limited to a particular zone then the orderly gradient of positional information could still be maintained throughout development by the activity of the cells at the boundaries. The results of rotation experiments in the goldfish suggest that the cells of the tectum, like those of the insect epidermis do have a partial memory and that this 'memory' is modified by an interaction with the cells around them.

This is necessarily a highly speculative comparison, but it emphasises the significance of boundaries in the organisation of development and it suggests experiments designed to locate and study boundaries within the central nervous system.

Gradients in the insect nervous system

In the insect, although the pattern of the integument varies from segment to segment, Locke (1959) discovered that the underlying gradient remains the same. This is true both of the body surface (Stumpf, 1968) and of the appendages (Tokunaga & Stern, 1965; Bohn, 1970; Postlethwait & Schneiderman, 1971) and Young (1972) has recently demonstrated that this serial homology in the epidermis is repeated within the central nervous system, and in the connections formed between the two. In his experiments he shows that there are homologous motor neurons in the meso- and metathoracic ganglia of the cockroach and that these neurons innervate equivalent muscles in the adjacent limbs. When a metathoracic leg is transplanted to a mesothoracic segment it retains its metathoracic characteristics during subsequent moults. Nevertheless, connections are established between the motor neurons of the mesothorax and their equivalent muscles in the transplanted metathoracic limb. At the same time, the sensilla of the transplanted leg establish central connections with appropriate interneurons as judged by the normal appearance of walking and the performance of levator reflexes.

Young concludes that these results show a difference between the segmental organisation of the nervous system and the epidermis. He suggests that the motor and sensory innervation may be independent of segmental gradients. Our interpretation is different. We feel that his results eloquently demonstrate that the segmental repetition of the gradient finds a counterpart within the central nervous system. The successful regeneration of the motor and sensory nerves implies that the information they require for patterned reconnection is repeated segment by segment and that they, like the cells of the metathoracic limb react to this homologous information in different ways, according to their segmental origins.

Our argument to this point has been indirect, in so far as it attempts to relate the characteristics of gradients – typified by the gradient in the insect segment – to the problem of positional information and polarity in the nervous system. However, the insect epidermis is itself supplied with numerous receptors and we draw attention here to the possibility that the gradient of positional information demonstrated by the regional differentiation of the integument also regulates the differentiation of the sensilla in the integument, as expressed by their connections within the central nervous system.

The density and the distribution of these peripheral sensilla depends on the occurrence of differentiative divisions among epidermal cells to form new receptor cell groups (Wigglesworth, 1953). Often, the distribution of these differentiative events is highly ordered and the arrangement of mechanoreceptor hairs in *Rhodnius* and *Drosophila* (Wigglesworth, 1940; Stern, 1968) is a model for the formation of patterns in the insect. There are few cases where the insect discriminates among the sensilla in such an array, but these examples are an opportunity to study the way in which neurons are identified centrally according to their position at the surface. Similar explorations of the mechanisms governing the formation of spatially identified connections in amphibia have been weakened by the central position of the afferent cell body. Such experiments allow contrasting conclusions, either central rearrangement in response to changes in peripheral 'sign', or alternatively, the specific reinnervation of skin grafts by central neurons uniquely appropriate to the type of skin concerned (Gaze,

4-2

1970). Insect material does not suffer from this ambiguity and the gradient theory provides a basis for assessing the class to which peripheral neurons are assigned (according to their spatial distribution) which is independent of their central connections.

Detailed investigations of the central connections of epidermal sensilla are only now beginning in insects (Callec, Guillet, Pichon & Boistel, 1971; Camhi, 1969*a*, *b*; Edwards & Palka, this volume). This information is required for it is at their junctions with higher order cells that the distinctive properties of otherwise similar receptors are revealed. Differences which are not reflected in the overt behaviour of the insect may nonetheless be present in a consistent pattern of connections within the CNS. The frontal hair sensilla of the locust are an example of a system where a property related to the surface position of the receptors regulates their endings within the CNS (although this information is not necessarily translated into an equivalent spatial array). Weis-Fogh (1949) identified these hairs as a control element in the flight machinery. Small changes in wind angle produce yaw-correcting postural changes and Camhi (1969*a*, *b*) has proposed an integrative model of the central endings which accounts for the directional coding which this stabilising behaviour implies. In the suboesophageal ganglion the hair sensilla are linked with cells which signal relative wind, wind direction, changes in wind angle and wind acceleration. In the case of the direction sensitive interneurons, each unit connects only with sensilla from a small part of the head. The directional sensitivities of these sensilla are similar, and since the summed activities of several are necessary to fire the higher order cell, the interneuron has considerably sharper directionality than any of the individual receptors with which it is connected. In this way directional information is coded by a selective connection of sensilla with central interneurons and the integrative mechanism originates in a recognition of differences between the axons of sensilla according to their position at the surface.

A second system has been described recently (Bate, 1972) in which a discrimination develops between sensilla at different levels in the same segment. It suggests that the distinctive property of the sensilla is their position in a segmental gradient. The system concerned is a simple defensive device known as the gin trap. Gin traps occur in the pupae of several different orders of insects (Hinton, 1955). They commonly consist of a series of dorsally or

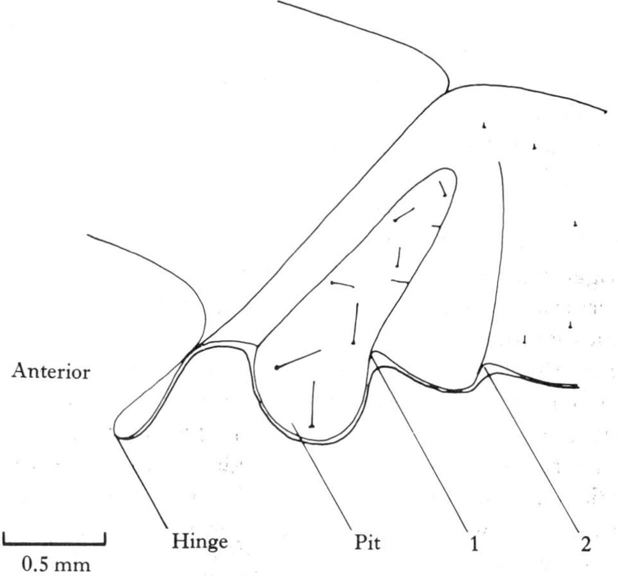

Fig. 2. A single gin trap from *Sphinx ligustri* seen in diagrammatic cross section. The function of the trap depends on a central discrimination between the hair receptors which lie inside and outside its margin. 1, 2, primary and secondary jaws.

laterally placed jaws at the anterior margins of the jointed abdominal segments, which are capable of crushing mites and small insects. These segmental traps (Fig. 2) are opened and closed by contractions of the persistent abdominal muscles which are triggered when the sensilla between the jaws are disturbed. The efficient trap closes only when objects move within it, so its function depends on a discrimination between the receptors which lie within its margins and others outside.

In *Sphinx ligustri*, where the pupa has three pairs of these traps, one on either side of the free abdominal segments, there is no difference between the response recorded peripherally from receptors inside the trap (which trigger closure) and those outside (which do not). Triggering therefore reveals a characteristic difference between the central connections of the two classes of receptors, both of which are derived from the abdominal sensilla of the preceding larval instar. At this earlier stage there is no functional difference in the central connections, so that the abdominal hair sensilla are divided into two classes by a differentiative process which occurs at pupation.

Connection with a segmentally arranged closure mechanism is a simple distinction within the central nervous system which roughly coincides with the division of the hairs at the surface of the animal into those inside and outside the traps. If these two classifications were to coincide exactly, then the criterion which distinguishes the hairs which connect with the closure mechanism would be that the triggering hairs lie within the margins of the developing trap. This criterion is of interest because it reveals the rules on which the classification depends.

When the fresh pupa emerges, its sensitivity to mechanical stimulation is so great that the disturbance of a single hair is sufficient to elicit the closure response. During this early stage, hairs connected with the closure mechanism can be separately identified and mapped. Maps prepared in this way (Fig. 3*b*) show that connection with the closure mechanism is not restricted to the hairs in the trap but includes other hairs, outside. The triggering and non-triggering hairs lie close together but there is a sharp distinction between them, so that hairs which fail to elicit closure when stimulated singly, also fail when they are stimulated in groups. The distribution of the two classes is quite consistent: the boundaries between them are always in the antero-posterior axis and never in the dorso-ventral. Clearly, connection with the closure mechanism is not confined to the hairs inside the gin trap but includes other hairs which lie outside the trap at a similar level in the antero-posterior axis.

During the development of the pupa, the differentiation of the abdominal segment produces a subdivision of the pupal cuticle which is similar to that studied by Stumpf (1968) in *Galleria*. She suggested that the diversification of the cuticle depended on an interaction between competent cells and a segmental gradient in the antero-posterior axis. In *Sphinx*, the subdivision of the abdominal segment coincides with a columnar arrangement of the epidermal cells which is revealed by an annulated cuticle which they secrete in the larva (Fig. 3*a*). A simple explanation for this association is that there are discrete differences between cells in adjacent columns which are incorporated into larger units at pupation. Notice that the gin trap and the sensilla within it are confined to a single column of epidermis considered in this way.

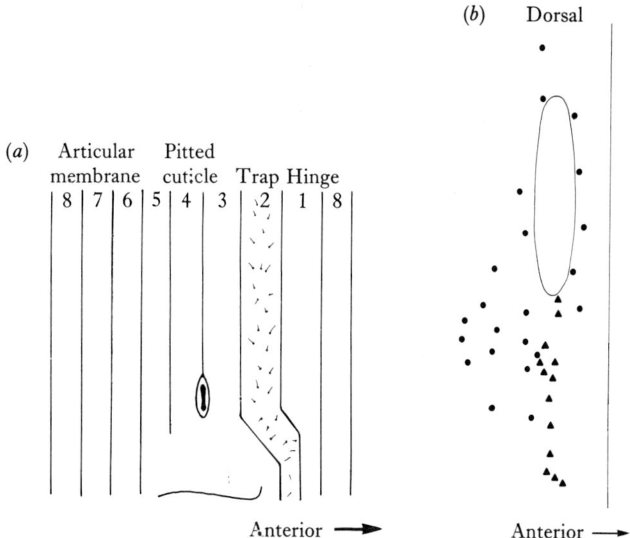

Fig. 3. (*a*) A map of the dorso-lateral region of a single abdominal segment in the fifth instar caterpillar, showing the division of the epidermis into columns and the regional differentiation of the pupal integument with which these divisions coincide. Mechanoreceptive hairs are uniformly distributed over the surface, but only those in column 2 are shown. (*b*) A map of the triggering (triangles) and non-triggering (circles) hairs in the region of the gin trap, shortly after the emergence of the pupa. All of the hairs inside the gin trap trigger closure and are not shown.

The trap is equipped with triggering sensilla by a combination of two separate processes: an incomplete degeneration of larval sensilla which leaves a strategic concentration of receptors within the trap and a selective connection of sensilla with a mechanism which closes it. The two processes fail to coincide precisely so that some hairs outside the trap also trigger closure. This departure from expectation suggests that the criterion which consigns persistent hairs to the triggering class is their position in the antero-posterior axis of the segment. More specifically the connection is restricted to those neurons whose cell bodies lie within certain levels of a segmental gradient on which the regional differentiation of the epidermis depends.

A consequence of the gradient theory is that positional information available to the cell body of the neuron must extend to the tips of its growing axon and dendritic branches. We have proposed, as a working hypothesis, that the positional information is encoded in the concentration of a substance which diffuses freely within the cell. *In principle* therefore, the value of the gradient is continuously available to the axon tip. In this view, connections depend, as do the distinctions between epidermal cells, on quantitative rather than qualitative differences between the developing neurons. Our view differs from comparable theories (Sperry, 1951) in which axial gradients establish qualitative differences between cells during embryogenesis. In a system of continuous variation there is a range of available connections open to the developing neuron and the connections which it finally makes depend not only on the terminals which are available to it but on a competitive interaction with other neighbouring cells.

We are grateful to Drs Mike Gaze and Clark Slater for their helpful comments on the manuscript.

References

Bate, C. M. (1972). Ph.D. thesis, University of Cambridge.

Bohn, H. (1970). Interkalare Regeneration und segmentale Gradienten bei den Extremitaten von *Leucophea* Larven (Blattaria). I. Femur und Tibia. *Wilhelm Roux Archiv für Entwicklungsmechanik der Organismen*, **165**, 303–41.

Burrows, M. (1973). Physiological and morphological properties of the metathoracic common inhibitory neurone of the locust. *Journal of Comparative Physiology*, **82**, 59–78.

Callec, J. J., Guillet, J. C., Pichon, Y. & Boistel, J. (1971). Further studies on synaptic transmission in insects. II. Relations between sensory information and its synaptic integration at the level of a single giant axon in the cockroach. *Journal of Experimental Biology*, **55**, 123–49.

Camhi, J. (1969*a*). Locust wind receptors. I. Transducer mechanics and sensory response. *Journal of Experimental Biology*, **50**, 335–48.

(1969*b*). Locust wind receptors. II. Interneurones in the cervical connective. *Journal of Experimental Biology*, **50**, 349–62.

Child, C. M. (1941). *Patterns and Problems of Development*. Chicago: University of Chicago Press.

Crick, F. H. C. (1971). The scale of pattern formation. *Symposium of the Society for Experimental Biology*, **25**, 429–38.

Gaze, R. M. (1970). *The Formation of Nerve Connections*. London and New York: Academic Press.

Gaze, R. M., Chung, S. H. & Keating, M. J. (1972). Development of the Retino-tectal Projection in *Xenopus*. *Nature New Biology, London*, **236**, 133–5.

Gaze, R. M. & Keating, M. J. (1972). The visual system and 'neuronal specificity'. *Nature, London*, **237**, 375–8.

Hinton, H. E. (1955). Protective devices of endopterygote pupae. *Transactions of the British Entomological Society*, **12**, 50–92.

Jacobson, M. (1970). *Developmental Neurobiology*. New York: Holt, Rinehart & Winston.

Lawrence, P. A. (1970). Polarity and patterns in the postembryonic development of insects. *Advances in Insect Physiology*, **7**, 197–260.

(1972). The development of spatial patterns in the integument of insects. In *Developmental Systems – Insects*, ed. S. J. Counce & C. H. Waddington. New York: Academic Press, in press.

Lawrence, P. A., Crick, F. H. C. & Munro, M. (1972). A gradient of positional information in an insect *Rhodnius*. *Journal of Cell Science*, **11**, 815–54.

Locke, M. (1959). The cuticular pattern in an insect *Rhodnius prolixus* Stål. *Journal of Experimental Biology*, **36**, 459–77.

Marcus, W. (1962). Untersuchen über die Polarität der Rumpfhaut der Schmetterlinge. *Wilhelm Roux Archiv für Entwicklungsmechanik der Organismen*, **154**, 56–102.

Postlethwait, J. H. & Schneiderman, H. A. (1971). Pattern formation and determination in the antenna of the homeotic mutant *Antennapedia* of *Drosophila melanogaster*. *Developmental Biology*, **25**, 606–40.

Sharma, S. C. & Gaze, R. M. (1971). The retinotopic organisation of visual responses from tectal reimplants in adult goldfish. *Archives Italiennes de Biologie*, **109**, 357–66.

Sperry, R. W. (1951). In *Handbook of Experimental Psychology*, ed. S. S. Stevens. New York: Wiley.

Stern, C. (1968). *Genetic Mosaics and Other Essays*. Cambridge, Massachusetts: Harvard University Press.

Straznicky, K. & Gaze, R. M. (1972). The development of the tectum in *Xenopus laevis*: an autoradiographic study. *Journal of Embryology and Experimental Morphology*, **28**, 87–115.

Stumpf, H. (1968). Further studies on gradient dependent diversification in the pupal cuticle of *Galleria mellonella*. *Journal of Experimental Biology*, **49**, 49–60.

Tokunaga, C. & Stern, C. (1965). The developmental anatomy of extra sex combs in *Drosophila melanogaster*. *Developmental Biology*, **11**, 50–81.

Weis-Fogh, T. (1949). An aerodynamic sense organ stimulating and regulating flight in locusts. *Nature, London*, **163**, 873–4.

Wigglesworth, V. B. (1940). Local and general factors in the development of 'pattern' in *Rhodnius prolixus* (Hemiptera). *Journal of Experimental Biology*, **17**, 180–200.

(1953). The origin of sensory neurones in an insect *Rhodnius prolixus* (Hemiptera). *Quarterly Journal of Microscopical Science*, **94**, 93–112.

Wolpert, L. (1971). Positional Information and Pattern Formation. *Current Topics in Developmental Biology*, **6**, 183–224.

Young, D. (1972). Specific re-innervation of limbs transplanted between segments in the cockroach *Periplaneta americana*. *Journal of Experimental Biology*, **57**, 305–16.

Development of the compound eye and optic lobe of insects

I. A. Meinertzhagen

Introduction

Just how the neurons of a nervous system form their functionally appropriate connections remains an important but largely unsolved problem still replete with vague and often conflicting ideas. This is in spite of a recent resurgence of interest represented in two recent reviews by Gaze (1970) and Jacobson (1970) (see the editor's introduction to this volume). The optic lobes of compound eyes, because of their spatially ordered morphology and the iterative nature of most of their connections, offer a system which is one of the least unfavourable for the study of the generation of specific neuronal connections. This review is an attempt to summarise present knowledge about the developmental processes of optic lobe involved in the fabrication of neuropile. Previous reviews of the development of compound eye and optic lobe have covered rather different aspects: development of the retina (Bodenstein, 1953), cellular proliferation of the optic lobes (Edwards, 1969; Panov, 1960), and the morphogenetic dependence between eye and optic lobe (Pflugfelder, 1958).

The emphasis throughout the review is to carry the analysis to single cell level, i.e. to analyse the behaviour of individual cells during the formation of their connections. In order to do this it is necessary to provide the developmental past history of the cells, by which their axons arrive at particular places in the partly formed neuropile, at particular times. Furthermore, the first essential is to know the exact sequence of development of the cells of the retina because it becomes apparent that much of the organisation of the optic lobe is imposed by the pattern of retinal development.

The stimulus for the single cell approach arises from detailed studies of the anatomy of insect optic lobes in general and in particular several contemporary studies of the fly optic lobe, notably by Trujillo-Cenóz and by Braitenberg's group at Tübingen. These studies have provided, in the first optic neuropile of the fly, what must be the most detailed knowledge of the connections in any neuropile. Part of the review will therefore be devoted to a special summary of recent observations on the development of this system. The review is both overdue and premature, many of the studies which it covers are either old and obscure or unpublished and preliminary.

The patterns of development of compound eye and optic lobe

Hemimetabolous larvae generally have functioning compound eyes like those of the adult, while holometabolous larvae are usually either blind or have lateral ocelli (stemmata).* These ocelli develop early on, independently of, but in association with the compound eyes.

The hemimetabolous type have small larval compound eyes which grow subsequently by either addition of new ommatidia or enlargement of existing ones (Bodenstein, 1953). The extent to which larval ommatidia are retained into the adult eye is however variable (Bodenstein, 1953). In paurometabolous forms generally the adult eye is the sum of its larval ommatidial increments. This is particularly obvious in various Orthoptera in which the instar additions are marked off by pigmented bands (Volkonsky, 1938). In dragonflies on the other hand, there is extensive duplication and overlap of ommatidial production in the nymphal and adult stages (Lew, 1933) and a reorganisation of the lamina at metamorphosis (Lerum 1968; Mouze, 1972).

In holometabolous insects the compound eye commences development after the first emergence of the larva and the major development occurs either in the larval stage as in Diptera and Hymenoptera or rapidly at the end of this stage and into the pupa as in some Coleoptera and Lepidoptera. In some of the Holometabola, e.g. Lepidoptera, Coleoptera, nematocerous Diptera, the eye forms externally, while in others it forms internally in specialised imaginal discs, invaginations of the larval pharyngeal epidermis (cyclorraphous Diptera and Hymenoptera). In some primitive Holometabola (culicomorph Diptera and hymenopteran sawflies for example) the compound eye functions in the larva although these forms have lateral ocelli which develop first and

* Compound eyes are also present in the larvae of endopterygote Mecoptera and the archaic lepidopteran Micropterygidae (Riek, 1970).

around which the developing compound eyes and their neuropiles are laid down. In larvae with lateral ocelli the ocellar (stemmatal) nerve connecting the larval ocellus or multiple ocelli to the larval optic ganglion serves as a guide for the first ingrowth of retinula axons of the compound eye. Although the ocelli may persist into the adult in a few forms they generally degenerate at metamorphosis and often are represented only by pigment spots in the head epithelium or on the lateral surface of the brain.

The wide variety of modes of compound eye and optic lobe development and their relationship to the development of larval optic centres, have conveniently been summarised by Panov (1960) in a way which illustrates the main trends. Despite this diversity in the overall development of the compound eye and optic lobes the cellular pattern of development is remarkably similar throughout the range in the formation of the retina and in the formation of the optic lobe neurons.

Cellular pattern of eye development

In general, eye development proceeds autonomously in sequences, first of cellular proliferation and determination, then of differentiation, which pass as waves across the prospective eye field, that region of the head epidermis which will transform itself into the compound eye. These processes are more easily observed in those forms in which the eyes develop externally as an integral part of the head epidermis. The subject has been previously reviewed by Bodenstein (1953, 1963).

The superficial manifestation of retinal differentiation proceeds as a sequence across the retina in either a postero-anterior or a postero-dorsal–antero-ventral direction, from a point in the posterior or dorsal part of the prospective eye region, which in some insects has been shown to function as a differentiation centre. This has been observed in various orthopterans (Jörschke, 1914; Volkonsky, 1938), in the cockroach *Periplaneta* (Hyde, 1972), in various dragonflies (Lew, 1933; Mouze, 1972; Schaller, 1960), in the hemipterans *Notonecta* (Lüdtke, 1940), *Corixa* and *Sigara* (Young, 1969) and *Rhodnius* (Laschat, 1944), in the lepidopterans *Bombyx* (Wolsky, 1947), *Ephestia* (Umbach, 1934) and *Doleschallia* (Burke, 1956), in the coleopterans *Gyrinus* (Bott, 1928), and *Dytiscus* (Günther, 1912), in a number of nematocerous Diptera (Constantineanu, 1930; Haas, 1956; Pflugfelder, 1937; Satô, 1951,

1953*a*, *b*; White, 1961; Zavřel, 1907) and in *Drosophila* (Becker, 1957; Gottschewski, 1960).

The mosquito *Aedes* has been especially closely analysed (White, 1961, 1963). The retina is formed by the activity of a development, or differentiation, centre near the posterior edge of the head capsule. This centre induces mitoses in the head epithelium which spread across the presumptive eye region. As mitotic activity increases, the epithelium thickens into an optic placode, the anlage of the retina, the advancing edge of which has the ability to induce further mitoses in the epithelium ahead of it.* As a result of these mitoses the cells become determined to form retina instead of epithelium. Although retinal determination normally spreads across the prospective eye region in one direction, in artificial situations it can be made to spread in many directions, i.e. there is no intrinsic polarity of transmission. White has concluded that the results are consistent with the intercellular diffusion of an inducing substance.

The differentiation centre is essential for the production of the eye, and destruction of the region of epithelium containing it prevents the appearance of the eye anlage and the later formation of ommatidia (White, 1963). This effect can be reversed by implantation of a fragment of the anlage into the head epithelium. Similarly Pflugfelder (1936–7) reports that, in *Pentatoma*, elimination of the anlage of the retina in the last two larval instars prevents regeneration of the retina. Similar conclusions have also been reached by Seidel (1935) for the dragonfly, by Wolsky (1956) for *Bombyx* and by Yagi & Koyama (1963) for various other Lepidoptera. In the cockroach, Hyde (1972) has shown that the retina does not regenerate if it is completely removed, but in the cricket *Acheta* a new eye is sometimes regenerated after heat cautery of the eye of early third instar nymphs (Heller & Edwards, 1968). It may be in the latter case that some remnant of a differentiation centre survived the cautery, but this seems unlikely because at the next moult the eye was totally replaced by apparently normal head cuticle.

In the mosquito, the competence of the head epithelium to

* By rotating a portion of this advancing edge with its surrounding head epidermis through 180° in the locust, Shelton (personal communication) has observed the production of supernumerary eyes which result from ommatidial induction in a retrograde direction.

respond to the inductive stimulus to form retina is limited to the cells of the prospective eye field. Head epithelium from other regions fails to produce ommatidia when placed in contact with the optic placode (White, 1961). In the cockroach however, developing ommatidia are able to induce the formation of additional ommatidia from prothoracic epidermal cells transplanted into the prospective eye region, or to induce them *in situ* when the eye is transplanted to the prothorax (Hyde, 1972). The phenotype of such an induced retina always expresses the genetic constitution of the original epidermal cells, rather than that of the inducing cells. In this insect then, the competence to respond to retinal induction is more widespread than in the mosquito, although within the head retinal induction is obviously still only effective in the area of the prospective eye region. Competence to respond to the stimulus of retinal induction may be restricted in a progressive manner which is more advanced in mosquito than cockroach at the stages to have been used for experimentation, or of greater interest, the competence to respond may be found in both head and prothoracic epithelia but only in the middle part or so of each segment. These are fascinating problems that deserve further attention.

The pattern of mitoses gradually forms ommatidial cell clusters, each with a final number of about twenty cells. Most insects have eight retinula cells and this is probably the primitive number. The cells of the dioptric apparatus usually separate early from those of the retinula and pigment cells but there is no definitive study on the lineage of the cell components of an ommatidium. The clearest account of the patterns of division is given by Bernard (1937), for *Formicina*. From this account is derived the cell lineage pattern by which one stem cell gives rise to all the cells of an ommatidium according to the scheme in Weber (1966). Whether the observations of Bernard have sufficient resolution to support such a lineage pattern decisively is equivocal,* although in view of its simplicity and close resemblance to the known lineage patterns of other epidermal structures, it is quite likely. Further observations on the ommatidia of Diptera are discussed later (see p. 69).

The mitotic activity of the eye field gradually gives way to waves of differentiation which pass across the newly formed ommatidial clusters. Frequently there is a temporal overlap between prolifera-

* Ommatidia of most Hymenoptera for example have a ninth basal retinula cell not included by Weber.

tion and differentiation in different parts of the eye. The cells begin to differentiate into the components of the adult ommatidium, often undergoing displacements in height within the ommatidium (for example in the sawfly – Corneli, 1924). A review of retinal differentiation will not be given here. Accounts are provided in many of the publications listed above dealing with individual species or insect groups. Patience and an aptitude for languages are necessary to extract the anatomical details, many of which need to be supplemented by electron microscopic observations.

The related topic of eye pigmentation, its genetic and humoral control are dealt with in the reviews by Bodenstein (1959) and Pflugfelder (1958).

Optic lobe development

The main proliferative events of the ganglion cell layers of the developing compound eye are now well established for a wide range of insect species, largely as a result of the comparative studies of Panov (1957, 1960) and the autoradiographic studies of Nordlander & Edwards (1969*a*, *b*; reviewed in Edwards, 1969) which have provided the first quantitative analysis.

The optic lobe comprises three major visual neuropiles which are, from retina towards the centre, lamina, medulla and lobula. (In Diptera and Lepidoptera the lobula is subdivided into lobula and lobula plate neuropiles.) Ganglion cells of these three optic lobe neuropiles develop from two optic anlagen, called variously proliferative centres, formation centres (*Bildungsherdes*), imaginal epithelia etc. The anlagen of each eye are derived from two separate lobes of the three which form the protocerebrum of the embryo (Heymons, 1895; Viallanes, 1891; Wheeler, 1893). The outer optic anlage which gives rise to the ganglion cells of lamina and medulla cortices is formed from the first, most lateral, of these three lobes (Bauer, 1904), while the inner optic anlage which gives rise to the cells of the lobula complex is derived from the lateral part of the second protocerebral lobe.

Unlike the anlage of the retina, which induces mitoses in a layer of pre-existing epithelial cells, the optic anlagen proliferate new ganglion cells into the regions nearby. As a result, large cell masses are formed, the shapes of which determine the early morphology of the developing optic lobe. Within these masses, those cells furthest from each anlage are oldest and are separated from it by a continuous, intervening population of cells with a spectrum of intermediate ages (Nordlander & Edwards, 1969*b*).

The anlagen contain aggregations of neuroblasts which divide either equally or unequally (Panov, 1957, 1960). Asymmetrical division yields from each neuroblast another neuroblast and a ganglion mother cell, while symmetrical divisions give rise only to additional neuroblasts. Whether neuroblasts divide according to one pattern or the other depends upon the insect species and the position of the neuroblast in the anlage, those undergoing symmetrical divisions being situated at the centre of the anlage (Nordlander & Edwards, 1969*b*; Panov, 1960). Ganglion mother cells subsequently divide and give rise to ganglion cells, although the number of intervening mitoses is still uncertain (Panov, 1960; Nordlander & Edwards, 1969*a*).

The control of proliferation and differentiation in the developing eye and optic lobe

A recurrent theme in developmental studies on eye and optic lobe, as in the comparable literature on the vertebrate visual system, is the dependence of differentiation between the two regions (Pflugfelder, 1958). In general the retina develops autonomously, whereas the optic lobe requires retinal innervation for normal differentiation.

Morphogenetic independence of retina from optic lobe

This problem has been reviewed at intervals (Bodenstein, 1953; Chevais, 1937; Edwards, 1969; Pflugfelder, 1958; Power, 1943). The evidence is still contradictory, the conflict apparently originating largely from a lack of attention to the stage of development at which operative procedures have been carried out.

Evidence *against* morphogenetic dependence of retinal differentiation upon optic lobe influence was found after optic lobe extirpation and eye transplantation in the moth *Lymantria* (Kopeć, 1922). A similar conclusion has been reached after eye transplantation in the cockroach *Periplaneta* (Wolbarsht, Wagner & Bodenstein, 1966), in the stick insect *Carausius* (Pflugfelder, 1947) and in *Drosophila* (Chevais, 1937; Steinberg, 1941; Vogt, 1946; Bodenstein, 1953). The following observations also support this conclusion. Optic lobe extirpation in the bug *Pentatoma* does not affect eye development (Pflugfelder, 1936–7). Eye differentiation is unaffected when the optic stalk fails to establish contact with the optic lobes in eyeless *ey*[2] *Drosophila* (Power, 1943) or after optic nerve section in the fly *Sarcophaga* (Schoeller, 1964). Finally, in the ant *Formicina* and the beetle *Tenebrio*, precocious retinula cell differentiation commences before the establishment of retinal innervation of the optic lobe (Bernard, 1937).

Evidence *for* a morphogenetic dependence of retinal differentiation upon an influence from the optic lobes has been found in *Periplaneta* (Drescher, 1960) and two species of Lepidoptera. Plagge (1936) found in the moth *Ephestia* that the eyes formed from transplanted eye discs lose their fine structure but that the effect could be eliminated by simultaneous implantation, with the eye disc, of part of the optic lobe to which it was attached. Wolsky (1938) found that cutting the (post-retinal) optic tract of various holometabolous insects causes abnormal development of the retina, the effects of which are localised mainly in the retinula cells. A further analysis (Wolsky & Wolsky, 1971) allows a provisional explanation of the conflicting evidence. The effect of optic lobectomy in pupae of the moth *Bombyx* is age dependent, the effect being greater for older pupae and greater for the older ommatidia of the posterior region of the eye. It seems most likely that ommatidia under some conditions merely undergo degenerative changes when their retinula axons are cut, as is also seen in third instar locust nymphs (C. M. Bate & J. Kien, personal communication). Ommatidial differentiation, according to this explanation, is independent of optic lobe influence. Results obtained from experiments employing optic lobe removal or eye disc transplantation will, however, depend upon the number of retinula axons already grown at the time of separation of eye disc and optic lobe, the susceptibility of ommatidia to degenerative changes, the efficiency of their recovery from these changes and the time available for recovery between axon section and emergence of the adult.

Morphogenetic interdependence of cell populations of the optic lobe and their dependence on the retina

What are the stimuli for proliferation and differentiation in optic lobe ganglion cells and how are these two distinct but continuous phenomena controlled and correlated? Most observations have been made on the effects of altered eye development upon the optic lobe, as manifest in the adult, they therefore mainly deal with the end product of development rather than the sequence of effects in development itself. In consequence interpretation is often unnecessarily difficult and would be greatly simplified if the developmental events had been studied directly and the various effects observed independently in their normal temporal sequence.

The control of neuroblast proliferation. In the adult, the cells of retina and optic lobe are present in fixed proportion. Therefore it is necessary to know if and how their production is controlled and co-ordinated.

The simplest situation to deal with is in the developing crustacean optic lobe, described by Elofsson & Dahl (1970). In the developing eye a semicircular proliferation zone situated medially buds off cells which are distributed to each of the retina, lamina and medulla externa (Fig. 5*a*). Possibly the proliferation centre contributes the cellular components, to each in turn, of an ommatidium in the retina and a cartridge in each of the lamina and medulla externa. There need be no problem involved in the control of the size of each individual population if one supposes that these proliferative activities are in some way linked. In the insect optic lobe the proliferative events of the retina are separate from those of the optic lobe and proliferation, at least within the outer optic anlage of *Carausius*, is independent of retinal innervation (Pflugfelder, 1947). Centripetal innervation may however exert a restricting influence on the rate of neuroblast proliferation. In the development of *Drosophila* wild-type, *bar* and eyeless ey^2 mutants, Hinke (1961) observed that the early growth rate of the whole optic lobe was greater in ey^2 than in *bar* which was greater than in the wild-type, while the contribution of neuropile growth to these growth rates was greater in wild-type than in either *bar* or ey^2. Apparently cell proliferation rates are initially faster in the mutants than in the wild-type. Within each of the optic anlagen the production of two ganglion cell populations could be linked in the same way as was suggested for the crustacean optic lobe. The different proliferation rates in lamina and medulla cortices observed in the butterfly *Danaus* (Nordlander & Edwards, 1969*b*) need merely represent the different numbers of ganglion cells which have to be produced for each cartridge of both regions.

There is an urgent need for an autoradiographic study of sufficient resolution to reveal the possible existence of linked proliferation into lamina and medulla cell cortices from the outer optic anlage and whether ganglion cells are produced as cartridge groups or as a uniform population which later becomes subdivided, without reference to clonal origin, by innervation. Alternatively genetic marking of the sort employed by Benzer and coworkers in mosaic *Drosophila* may be a more promising approach for the clonal analysis of the products of anlage mitoses. A study of this sort would also reveal whether, within a single cartridge group, determination of specific morphological cell types is transmitted by lineage or acquired in some other way.

Direction of axon outgrowth. Ganglion cells arise from ganglion mother cell mitoses with consistent orientations to the neuropile (Nordlander & Edwards, 1969*a*; Panov, 1960). It would be easy to imagine that this confers polarity on the cells which determines the direction of axon outgrowth during differentiation. In lamina monopolar neurons of the butterfly *Pieris*, however, axon outgrowth is not consistent, but in a variety of directions, often initially centrifugal but later reversing to become centripetal (Fig. 8*a*) (Sánchez, 1919). These varieties are also seen in the adult of *Pieris* (Fig. 7) (Strausfeld & Blest, 1970) and apparently do not correlate with the type of monopolar neuron, whereas in the fly the axon usually emerges from the proximal surface of its cell body (Strausfeld, 1971*a*).

The onset of axon outgrowth. A common feature in the development of the optic neuropiles is the centripetal (i.e. from the retina inwards) innervation of each optic neuropile, the cells of which send out axons in a sequence across the neuropile to grow to the next via the chiasmata. What is the stimulus for the onset of axon outgrowth and why does it occur in a sequence across the neuropiles? More specifically, is the stimulus for ganglion cells to differentiate axons intrinsic to a population of cells or is it triggered extrinsically by innervation from an adjacent layer? There is evidence to support both alternatives which, because of their simpler connections and direct innervation from the retina, is clearest in the case of the cells of the lamina and medulla, i.e. those produced by the outer optic anlage. In some insects the evidence supports a mechanism in which axon outgrowth is dependent upon retinal innervation, while in others that it is independent of innervation.

To accept that *axon outgrowth depends upon retinal innervation* first requires the observation that innervation of the lamina and medulla by retinula cell axons precedes axon growth from the cells of either of these neuropiles. Direct observation of this sequence has been made only in *Pieris* and the dragonfly *Agrion*. The growth pattern is the same in both; long retinula axons leave the retina and go first to the medulla while short retinula, lamina monopolar and (in *Pieris* at least) transmedullary axon growth follows later. This evidence in *Pieris* conflicts with that from other Lepidoptera (see p. 64). In *Agrion* (Richard & Gaudin, 1960), where the lamina has not yet appeared at the time of emergence from the egg, the evidence is perhaps the clearest available because the larva of this genus on hatching has only seven ommatidia (Ando, 1957) which have grown retinula axons to the medulla.* The evidence from other dragonflies, however, does not provide decisive support for this observation (Lerum, 1968; Mouze, 1972). For example it appears from Mouze's figure of the first instar larval optic lobe of *Aeschna* that the lamina and medulla neuropiles develop before the first retinula fibres innervate them. Although different dragonflies may have different modes of development, it is more likely that newly grown retinula axons which are difficult to see, even with the best reduced silver stains, have been overlooked except in favourable cases like that of *Agrion*. In other insects, e.g. *Culex* (Pflugfelder, 1937) and *Apis* (Panov, 1960), the most that can be said is that lamina monopolar axon growth starts at about the same time as the first incoming retinula axons but has not been directly observed to precede it.

In the flies *Calliphora* and *Lucilia* retinal innervation also appears to precede axon outgrowth from cells of the lamina or medulla (Meinertzhagen, unpublished). Lamina monopolar ganglion cells only ever stain by reduced silver methods in that part of the lamina field already innervated by the retina (Fig. 3). Furthermore, in electron micrographs at the edge of the developing lamina, no isolated bundles of monopolar axons have been seen without accompanying long retinula axons, which characteristically have

* Long retinula axons were in fact not mentioned by Zawarzin (1913) in his original description of dragonfly optic lobe, but there are reasons to suppose that they exist in this insect as in others (Strausfeld & Blest, 1970).

an electron-dense axoplasm. Obviously these observations on lamina monopolar cells need to be checked carefully and extended also to the growing transmedullary and medulla centrifugal cells. A developmental relationship between centripetal lamina, medulla and retinal elements in *Drosophila* has been pointed out by Hanson (1972) because the positions of lamina and medulla cortices at the growing edge of the lamina plexus allow the simultaneous but temporary contact of retinula axons with lamina monopolar and medullary cortex cells (Fig. 3) which suggests that axon sprouting occurs in both in response to centripetal innervation.

To prove that axon outgrowth is induced by the ingrowing retinal innervation however requires that if the retinal innervation is withheld, lamina and medulla cells fail completely to differentiate axons. This evidence is available in insects which have had their retinal innervation eliminated either by genetic mutation or surgical interruption. Taken as a whole the evidence is fairly conclusive, at least in *Drosophila*, but no single report has been made with sufficient precision to be decisive.

Substantial evidence comes from the early investigations of Power (1943) subsequently confirmed by Hinke (1961) which sought to evaluate the effect of eye reduction factors in *Drosophila* upon the optic lobe neuropiles. In his study of the series wild-type, *bar* and eyeless ey^2 mutants in which the ommatidial field size is reduced progressively from normal to complete eyelessness, Power measured the volumetric hypoplasia of the optic neuropiles associated with each decrement of retina. With the total absence of ommatidia this is 100% (lamina), 85% (medulla), 59% (lobula) and 57% (lobula plate). His well known conclusion was that the hypoplasia in each neuropile was proportional to the extent that the neuropile received centripetal innervation, i.e. probably resulted directly from the reduced ingrowth of centripetal fibres in varying degrees of eye reduction. This is in conflict with evidence from *in situ* section of the optic stalk in *Sarcophaga*, where Schoeller (1964) found complete absence of the optic lobe neuropiles on the operated side. It appears at least possible from her figure (Plate IV, 3) that she failed to observe the 'missing' optic lobe which was probably situated in the thorax. It is known (Shatoury, 1963) that continuity between retina and brain is necessary to draw the optic lobe into the cephalic region during pupal eversion of the eye disc (see p. 73).

The optic neuropile growth rates of wild-type, *bar* and ey^2 mutant *Drosophila* have been studied by Hinke (1961), who found that the curves for wild-type and *bar* are qualitatively similar but the values for *bar* are lower, indicating that the same developmental events occur in both but in smaller number in *bar*. In the medulla of ey^2, the tiny adult neuropile volume is attained quickly, soon after pupation, indicating that the lobula cells have already sent their centrifugal processes into this neuropile. Therefore there can be no transient innervation of the medulla from the lamina which later regresses, because the absolute medulla neuropile volume never diminishes from its peak at pupation.

From these observations it can be deduced that, at least in the monopolar cells of the fly lamina, *virgin ganglion cells of the outer optic anlage do not grow axons*. This raises the important point that not even differentiation to the point of axon growth is an automatic outcome of a ganglion cell's genesis but needs to be initiated. Failure of the cells of the outer optic anlage to grow axons would deprive the neuropiles of the lobula complex of centripetal innervation, but in eyelessness nearly half the neuropile volume remains. The cells of these neuropiles produced by the inner optic anlage therefore probably put out axons in the absence of this centripetal innervation. The stimulus for axon growth in these cells may be intrinsic or conceivably derived from their central innervation through connections with the brain.

Optic lobes show reduced neuropiles in several insects if their retinal innervation is withheld completely before it even grows or if it is eliminated some time after the first retinula axons have grown.* The distinction between optic lobe cell degeneration and failure to differentiate following removal of the retina can only be drawn if the extent of retinal innervation is known at the time of the operation. Two findings directly confirm those of Power (1943) on *Drosophila*. In *Lymantria* (Kopeć, 1922) when the eye anlage was removed from the fourth instar larva, before retinal differentiation, the lamina and external chiasma were absent and the medulla and internal chiasma 'often but slightly developed'. In *Carausius*, Pflugfelder (1947) reports similar observations after transplanting optic anlagen.

A slightly different question is whether the optic neuropiles

* According to Holmgren (1909) there are no optic lobes in the blind castes of the termite *Eutermes*.

degenerate if their previously established retinal innervation is destroyed. Alverdes (1924) found that in *Cleon*, *Agrion* and *Notonecta* the lamina was absent from all, together with the medulla (*Cleon*), and some further degeneration in the lobula (*Notonecta*), after cautery of the larval retina. These experiments have been repeated rather more carefully by Stein (1954) in young dragonfly larvae. She confirmed that degenerative changes occur in both lamina and medulla but attributed the variability in the extent of degeneration found by Alverdes to the regenerative changes which commence in dragonfly after the first 14 days. Similarly Heller & Edwards (1968) found that the volumes of lamina and medulla are smaller if retinal reinnervation fails, while if it succeeds their organisation only is affected. Recent electron microscopic evidence from the adult fly *Musca* clearly reveals transsynaptic degenerative changes after localised retinula axon degeneration (Campos-Ortega & Strausfeld, 1972). On the other hand, Pflugfelder (1936–7) has shown by careful destruction of ommatidia in the last two larval instars of *Pentatoma* that the consequent retinula axon degeneration does not result in degeneration of the underlying neuropiles. Further work is necessary to clarify this point of conflict.

Proof that *axon outgrowth is independent of retinal innervation* requires the observation that either axon outgrowth from lamina and medulla cells precedes their innervation from the retina or that if retinal innervation is withheld from the beginning, axon outgrowth still proceeds.

Evidence of the first type is that in *Danaus* cells of the medulla cortex start their axon growth in the second instar, although the first lamina fibres to appear cannot be discerned until the end of the fifth larval instar (Nordlander & Edwards, 1969*b*). Furthermore the eye imaginal disc is not connected with the optic lobe until the optic neuropiles are well advanced, and conspicuous retinal innervation is not apparent until fully two days after pupation. This account is fundamentally different to that in the fly and the other insects already mentioned.

Evidence of the second type comes from the work of Schrader (1938) who transplanted the brain of fourth and sixth instar larvae of *Ephestia* into the abdomens of hosts of the same age. The transplants developed optic lobes differing little from normal. Schrader concluded that development of the optic lobe is autonomous from the fourth instar larva onwards in spite of the first retinal innervation not occurring until about the time of pupation (Umbach, 1934). Although their shape was rounder than normal, the implanted optic lobes contained apparently normal medulla and lobulae. The lamina on the other hand developed into a structureless ball from which Schrader concluded that the eye is necessary to determine its size and structure although not its presence. Analysis of the results would have been greatly facilitated by observations of sectioned material rather than implant whole mounts, but the evidence obtained by this author shows that the cells of the outer optic anlage can differentiate without ever receiving their normal retinal innervation.

These two pieces of evidence provide support for an alternative mechanism of initiation of axon outgrowth in the cells of the outer optic anlage and for the role of cell death in the restriction of the cell populations. Isolated degenerating ganglion cells have been observed in the developing optic lobes of *Danaus* (Nordlander & Edwards, 1968) which have been interpreted as death of surplus ganglion cells which fail to establish a 'peripheral' innervation. The normal sequence of events described by these authors involves a small but consistent over-production of ganglion cells and the subsequent cell death of those remaining uninnervated. This scheme does not exclude the possibility that every cell to be produced inevitably differentiates an axon, but does exclude retinal innervation as the initiating stimulus for axon growth, which must reside in some intrinsic mechanism.

It is important that subsequent work attempts to provide evidence to distinguish between these two mechanisms. For the next advance the evidence must provide details of the first appearance of specific cell types, rather than regions of fibre connection. Alternatively, information on the first appearance of specific neuropile layers which may be correlated with growth of the processes of known cell types is needed, rather than the first appearance or volumetric growth of the whole neuropile, which are ill-defined quantities.

I. A. Meinertzhagen

The development of eye and optic lobe in the fly

Development of the eye disc

Early studies on eye development in *Drosophila* have been reviewed by Bodenstein (1950). The eye-antennal discs arise as frontal sacs, hypodermal evaginations of the dorsal wall of the pharynx. By culturing fragments of the discs in the abdomens of host larvae Ouweneel (1970) has extended earlier observations by Vogt (1946) and constructed a fate map of the posterior portion of the wild-type right eye-antennal disc of *Drosophila* (Fig. 1*a*). The presumptive eye region is located in the flat central part of this disc (Bodenstein, 1950; Gottschewski, 1960; Ouweneel, 1970).

By analysing the size of mosaic spots in adult eyes of *Drosophila*, produced by X-ray-induced somatic crossing over in different aged larvae, Becker (1957) demonstrated the clonal nature of proliferation in the eye disc.

At the end of the first larval instar the anlage of the retina consists of about twenty cells, each of which gives rise to about forty ommatidia. At this time the orientation of the plane of cell divisions changes from a direction antero-dorsal–postero-ventral (for the ventral half) to that perpendicular to it and each of the twenty or so cells gives rise to a strip of ommatidia (Fig. 1*b*, *c*) reminiscent of the clonal patterns observed in many epidermal disc structures of *Drosophila* (Postlethwait & Schneiderman, 1971). In muscoid flies no comparable analysis is yet available but the numbers of ommatidia produced are about five times as great as in *Drosophila*. Furthermore the array of the middle and posterior ommatidia is distorted as if subject to a horizontal stretching from anterior to posterior (Braitenberg, 1967, 1970).

Cell division in the imaginal eye disc is reported to cease in the mid-third instar larva of *Drosophila* (Waddington & Perry, 1960) although Schoeller (1964) reports mitoses at the periphery of the eye disc up to fifty-four hours after pupation in *Calliphora*. In the pre-evaginated pupal eye disc, degeneration of some of the cells is indicated by pycnotic nuclei which are sometimes fairly numerous (Schoeller, 1964). This is the only evidence for an eliminative phase in retinal development, all other studies having exclusively revealed the progressive expansion of cell numbers up to the limits set by the determination of the retinal field.

From the third instar larva onwards, the presumptive retina

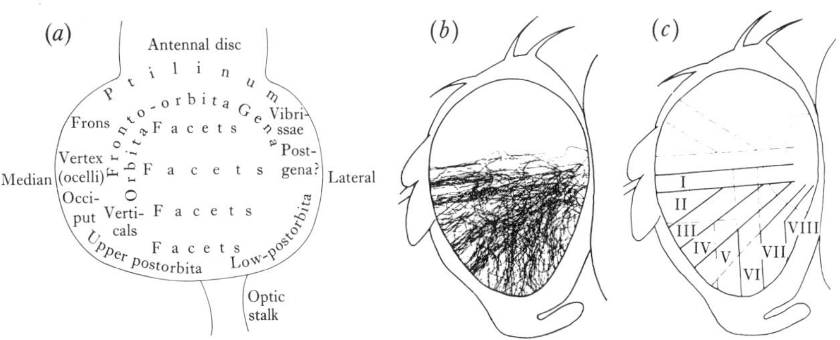

Fig. 1. (*a*) Diagram of the tentative fate map of a right eye disc of *Drosophila*. The folded borders of the disc give rise to the dorsal ocelli and the named parts of the head capsule, while the central part of the disc gives rise to the facetted eye region (From Ouweneel, 1970).

(*b*) Superimposed outlines of the pigment mosaic spots of the ventral eye half resulting from somatic cross over after X-ray treatment of embryos and first instar larvae of *Drosophila*.

(*c*) Schematic plan of the mosaic spot profiles reproduced in (*b*), showing the most commonly encountered shapes, sizes and positions in the ventral eye half (solid lines) and also those less commonly encountered in the ventral and equatorial eye regions (dashed lines). The dotted semicircular area at the midposterior eye border is inferred to be a region of minimal mitotic activity from the shapes of the clonal strips I–VIII which radiate away from it. Cells furthest from this region give rise to larger number of ommatidia. From the size of the mosaic spots produced by X-ray treatment at different stages, the strips I–VIII are inferred to be the mitotic products of single cells in the eye anlage at the end of the first instar larva. At this stage determination of the ommatidial field occurs and each of the 20 or so cells of the presumptive eye region becomes determined to produce a special part of the adult eye (From Becker, 1957).

contains numerous ommatidial clusters, visible by light microscopy, which probably contain the final complement of ommatidial cells. Each cluster must contain eight retinula cells because electron micrographs at this stage clearly reveal ommatidial bundles of eight retinula axons (Meinertzhagen, unpublished; Waddington & Perry, 1960). The composition of these clusters, even the sequence of cluster formation is not readily discernible by light microscopy and all the early light microscopical studies of eye discs at this stage are necessarily inadequate.

The ommatidial clusters which earlier are packed very closely at staggered depths in the retinal epithelium are drawn out into a single layer soon after eye disc eversion in the young pupa (Waddington & Perry, 1960). After eversion they undergo rapid growth and lengthen until they attain the elongate cylindrical form of the adult ommatidium (Waddington & Perry, 1960; Perry, 1968). Further details of the differentiation of the ommatidium of *Drosophila* are given by Waddington & Perry (1960), Waddington (1962) and Perry (1968) while the arrangement of cells within the ommatidium of *Calliphora* in the pupa is given by Schoeller (1964).

Differentiation commences as a wave which spreads anteriorly from the posterior border of the eye disc (Becker, 1957; Gottschewski, 1960). Probably axon growth starts synchronously in each vertical ommatidial row as the wave of differentiation passes (see caption to Fig. 2). Separate bundles of eight retinula axons emerge through the basement membrane on the inner surface of the retinal epithelium and enter the neck of the optic stalk. They continue to grow from the eye disc up till the early pupa.

Origin of the equator and the differences between dorsal and ventral halves of the eye.*

Although the cell lineages of dorsal and ventral ommatidia are separate from the end of the first larval instar or earlier (Becker, 1957), the borderline between the two is not coincident with the equator of the eye as defined by ommatidial rhabdomere patterns. This has been shown by mosaic analysis also in *Drosophila* in which clonally transmitted mosaic patches with various structural and pigmentation abnormalities were studied in regions near the equator (Hanson, Ready & Benzer, 1972). Many such mosaics were found to contain both dorsal and ventral rhabdomere patterns. This means that the position of the equator is not transferred clonally but must be established much later on. The establishment of the equator need not of course occur all at one time, but may unroll progressively as the extent of cellular proliferation in any part of the eye disc reaches the required stage. Culture experiments of eye disc fragments which could reveal the temporal pattern of the determination of the equator are badly needed. Presumably the late determination of the equator in retinal proliferation reflects the late acquisition in Diptera of a retinal organisation based on

* Details of the adult organisation of retina and optic lobe are briefly summarised in the caption to Fig. 9.

dorso-ventral mirror-image symmetry of radially asymmetrical ommatidia.

The production of the radially asymmetrical ommatidia of brachycerous flies could be by either an asymmetrical pattern of mitoses from a single stem cell or an asymmetrical pattern of aggregation of cells derived from more than one stem cell. Initial evidence (Hanson, Ready & Benzer, 1972) using mosaic tissue with retinula cell pigmentation markers indicates that a clonal origin is not involved in establishing the retinula cell types or their asymmetrical patterns within the ommatidia, while the close correspondence of the distribution of marked retinula and pigment cells at the edges of mosaic patches indicates that extensive cell migration between these two cell types does not occur. The ommatidia therefore seem to arise from asymmetrical patterns of cell aggregation of neighbouring cells, but the nature of these patterns is not yet clear. Haas (1956) also suggests that a phase of aggregation occurs during ommatidial production in the mosquito *Culex*.

Morphogenesis of the optic stalk

Early in larval development the eye discs contact the surfaces of the larval hemisphere and continuity is established between the two by the formation of the optic stalk (Fig. 2). Apparently the eye disc only forms the optic stalk normally when its base is correctly positioned in relation to the hemisphere (Gottschewski, 1960) although in some eyeless ey^2 mutant larvae the optic stalk is dislocated and connects the hemisphere with the anterior edge of the disc at its boundary with the antennal discs (Shatoury, 1963).

The optic stalk is first found at the beginning of the second instar larva or slightly before (Steinberg, 1941) but the first retinal fibres do not grow until nearly half way through the third instar larva in *Lucilia* and *Calliphora* (Meinertzhagen, unpublished) and *Drosophila* (Hanson, 1972).

The optic stalk of a mid-third instar larva consists of a substantial outer cellular sheath with a core of axons. The axons are separated into two groups, the 'ocellar' axon bundle and the retinula cell axons.

The 'ocellar' axon bundle. This consists of a few fibres (27 in *Lucilia cuprina*, 50 in *Calliphora stygia*) and is found in the optic

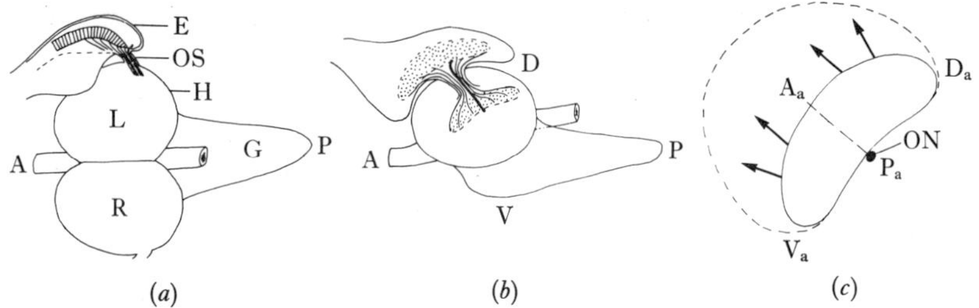

Fig. 2. Diagrams of the brain of a late third instar fly larva: (*a*) dorsal view, (*b*) lateral view of the left eye disc and hemisphere. The eye imaginal disc E is connected to the supraoesophageal hemisphere H through the optic stalk OS. The first retinula axons grow between the posterior margin of the eye disc and the posterior region of the larval hemisphere. The growth of axons is symmetrical from dorsal and ventral halves of the retina. (*c*) Enlarged view of the innervated area of lamina cortex. The retina–lamina projection appears in cross section as a growing ellipse of retinal axon bundles with an unchanging posterior margin indicated by the position of the 'ocellar' nerve bundle ON. Growth is achieved by the increment of axon bundles symmetrically along an anteriorly advancing front (indicated by arrows). This results in a slow postero-anterior sequence of axon arrival at the lamina which presumably reflects the wave of retinal differentiation and axon outgrowth in the eye disc and probably also rapid equator–dorsal and equator–ventral sequences which may simply reflect the greater distance that non-equatorial axons have to grow. A very approximate estimate of the interommatidial time lag for arrival of retinula axon bundles at the lamina may be derived in the following way, assuming an estimate of 25 μm/h for the growth rate of retinula axons (at 26 °C in *Lucilia*, see p. 71). If the postero-anterior sequence of axon arrival at the lamina occurs at a constant speed which (in *Calliphora*) takes approximately three days for completion of the growth of fifty rows then new axon bundle rows would be added at the rate of 1 per 90 min, ignoring smaller differences resulting from different path lengths. If the vertical sequence of axon arrival at the lamina results from the greater path lengths of dorsal or ventral axons (approximately 150 μm longer than equatorial axons in *Calliphora*) then a 6 h time lag would exist between axons at either the dorsal or ventral margins and the equator. New axon bundles would be added within each dorso-ventral row (of 100 bundles in *Calliphora*) at the rate of 50 in 6 h or 1 per 7 min, ignoring differences resulting from different path lengths in the eye disc which would reinforce the time lag. The consequence of these growth processes is that at any one time only a few dorso-ventral rows are adding new bundles and these are added at widely separated points. A, anterior; P, posterior; L, left; R, right; D, dorsal; V, ventral; G, suboesophageal ganglion; A_a, P_a, D_a, V_a, the corresponding axes of the adult optic lobe which have an oblique and variable orientation to those of the larval brain.

stalk from the second instar onwards, long before the first retinula axons. It arises from the posterior part of the eye disc but the cells of origin have not been identified. It apparently corresponds to the ocellar nerve of *Culex* (Pflugfelder, 1937) in which the larval ocellus develops early on and functions from the first instar as the first larval organ of sight. That the bundle is truly homologous with the nerve of the larval ocellus of *Culex* is suggested by its position and early appearance preceding the growth of the retinula axons, and its origin in a group of cells at the posterior border of the eye disc. Yet these cells have no contact with the outside nor have any ultrastructural specialisations, such as might be expected of a photoreceptor, been seen although these would be conspicuous amongst the undifferentiated cells in this region.

If this is the 'ocellar' nerve we are apparently presented with a curious developmental vestige. Although visually functionless it continues to play its primitive developmental role of sending pathfinder axons down the optic stalk and into the brain, preceding the retinula axons of the compound eye.

After entering the larval hemisphere the 'ocellar' bundle runs along the mid-posterior border of the lamina and is situated along the surface of the medulla neuropile (Fig. 3). The bundle is critical in the establishment of the optic neuropiles of the compound eye because the first axon pathways of this system are laid down along it, but its persistence into the adult is perhaps doubtful.

The retinula axons. Although the first few retinula axons of the compound eye run down the edge of the 'ocellar' nerve they are distinguishable by their smaller size and the two groups of axons never mix. Very shortly after, the axon groups are clearly separate and the retinula axons which are found in the optic stalk of the third instar larva of *Lucilia* from approximately 36 h onwards increase in number to occupy most of the area of the optic stalk at pupation (75 h). The retinula axons soon become split into their original (ommatidial) bundles of eight by glial invasion from the eye disc.

In *Lucilia* the first retinula axon bundles take approximately four hours to grow between eye disc and hemisphere at 26 °C (Meinertzhagen, unpublished) so that their growth rate is about 25 μm/h. In consequence no bundle can have completed its journey to the lamina before neighbouring bundles also start to

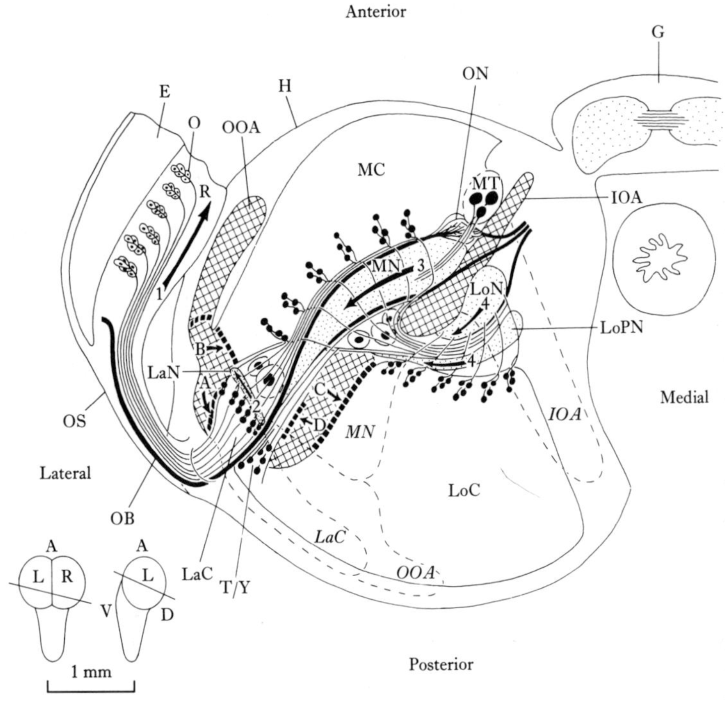

Fig. 3. Diagram of a section of the left hemisphere of the larval brain of *Calliphora stygia* at pupation as seen in reduced silver preparations. The plane of section (inset, left) is selected to cut the 'ocellar' nerve longitudinally, i.e. to contain the antero-posterior axis of the developing optic lobe. The section is shown from the dorsal surface and includes the first dorsal stratum of the external chiasma. Arrows (1–4) indicate the direction of growth by the addition of new axons in the retina, lamina plexus, medulla and lobula neuropiles. Two important inter-neuropilar axon growth regions occur at the advancing edges of the lamina and lobula plate neuropiles. At the first, retinula axons are in the vicinity of lamina monopolar and medulla cortical cells. At the second, transmedullary axons are in the vicinity of the different cell types of the lobula cortex. Heavy interrupted lines separate neuroblast aggregations from the ganglion cells to which they give rise. Arrows (A, B, C, D) in the outer and inner optic anlagen indicate the direction of proliferation of new ganglion cells into lamina, medulla and lobula cortices. Two monopolar ganglion cells are illustrated for each lamina axon bundle, out of the five found in the adult. Only four ganglion cells each are illustrated for the axon bundles of the medulla and lobula plate neuropiles of many found in the adult. The medulla cortex contains lamina centrifugal and transmedullary cells and those Y cells with a transmedullary cell body fibre. The lobula cortex contains T cells and Y cells, those T cells with cell body fibres which cross the lobula plate neuropile are proliferated in a sequence from C, the remaining T cells and Y cells are probably proliferated in a sequence from D.

grow alongside it, if the entire retinula projection is to be completed by the early pupa.

After pupal eversion of the eye disc the optic stalk undergoes a remarkable shortening by which the brain is pulled into the lumen of the eye disc (Shatoury, 1956). In some mutant ey^2 *Drosophila* in which the optic stalk fails to connect between eye disc and brain, the brain becomes separated from the retina by the cephalothoracic constriction and in the adult is found in the thorax (Shatoury, 1963). The optic stalk, which in larval development is long and narrow, becomes short and wide in the pupa and in the adult is represented by the axons and glia in the narrow zone between the basement membrane of the retina and the ganglion cell zone of the lamina (Gieryng, 1965; Meinertzhagen, unpublished).

Proliferative growth of the optic anlagen

The proliferative sequences are already reported in a general way for Diptera, in *Culex* (Pflugfelder, 1937). The visual neuropiles

---

The cell types formed by regions A–D are largely inferred by the positions of their cell bodies as given by Strausfeld (1970). Large cells, conspicuous in the external and internal chiasmata, are illustrated, although smaller (lamina epithelial and amacrine ?) cells are also found in the external chiasma. The large cells possibly include two types of tangential neuron of the lamina and medulla (lam:tan 1, M:tan 7 respectively; Strausfeld, 1970). The medulla tangential cells are however mainly found in a group at the medial edge of the medulla neuropile, their cell bodies are larger than those of the other ganglion cells and their axons can be seen entering the medulla in the position illustrated. A portion of the eye disc is shown. It lies outside the plane of the section anteriorly and dorsally. The ventral extents of the optic anlagen, lamina cortex and medulla neuropile are indicated by interrupted lines and identified with italicised letters. The cells of origin of the 'ocellar' axon bundle are not shown nor is the first ocellar neuropile which in *Culex* lies between lamina and the advancing edge of the medulla (Pflugfelder, 1937) but which is not obvious in *Calliphora*. The 'ocellar neuropile' in the diagram, which by its position corresponds to the second ocellar neuropile of *Culex*, is a conspicuous feature of the larval hemisphere because, being older, it has a higher silver affinity than the optic neuropiles surrounding it. E, eye disc; G, suboesophageal ganglion; H, supraoesophageal larval hemisphere; IOA, inner optic anlage; LaN, lamina neuropile (plexus); LaC, lamina cortex; LoC, lobula cortex; LoN, lobula neuropile; LoPN, lobula plate neuropile; MC, medulla cortex; MN, medulla neuropile; MT, medulla tangential cells; O, ommatidial cell cluster; OB, 'ocellar' axon bundle; ON, 'ocellar' neuropile; OOA, outer optic anlage; OS, optic stalk; R, retina; T/Y, T cells and Y cells.

arise from ganglion cells which are proliferated from two optic anlagen (Hertweck, 1931; Viallanes, 1885). No exact study of the proliferative dynamics or clonal patterns of the ganglion cells is yet available for the fly optic lobe, although this is a most favourable system for such an analysis, at least in the lamina, because of the amount of detailed information on neuron cell types and the correlation with cell body positions. Neuroanatomists concentrating on the adult structure of the optic lobe distinguish a wide variety of cell types by their Golgi-impregnated profiles (Cajal & Sánchez, 1915; Strausfeld, 1970; Strausfeld & Blest, 1970) which are classified according to their principle axis and the geographical extent of their processes. From a developmental standpoint a more fundamental distinction is between periodic and aperiodic elements with, presumably, early separation of the cell lineages of the two types.

Periodic ganglion cells are added orthogonally from the outer anlage to the lamina and medulla cortices. Proliferation in the inner optic anlage seems to be in two parallel directions (Fig. 3) the first producing T-cells and some Y-cells, the second in the body of the lobula cortex, producing the remaining T-cells. Further details are given in the caption to Fig. 3.

The production of aperiodic optic lobe elements is clear only in the case of those medulla tangential cells which are the first to be proliferated from the outer optic anlage into the medulla cortex, as has been shown decisively in *Danaus* (Nordlander & Edwards, 1969*b*).

Plate 1. The appearance of the lamina in (*a*) 15-hour (*b*) 2-day (*c*) 4-day and (*d*) 5-day pupae of *Calliphora stygia*. In the two earlier stages the lamina is represented at the internal face of the ganglion cell zone G by the lamina plexus L which later becomes transformed into the external plexiform layer of the lamina neuropile. The formation of cartridges by the centripetal growth of retinula terminals (arrowed in *c* and *d*) has occurred in the two later stages (i.e. approximately half way through pupal life) although individual axon connections comparable to those of the adult are not apparent until the oldest stage shown (*d*). The chiasma C is at first the narrow axon band between lamina plexus and medulla neuropile M in (*a*). Later it elongates and its axon bundles become widely separated by large glial expansions which give this region a vacuolar appearance in later pupae. 10 μm wax sections stained by Fraser-Rowell's reduced silver method.

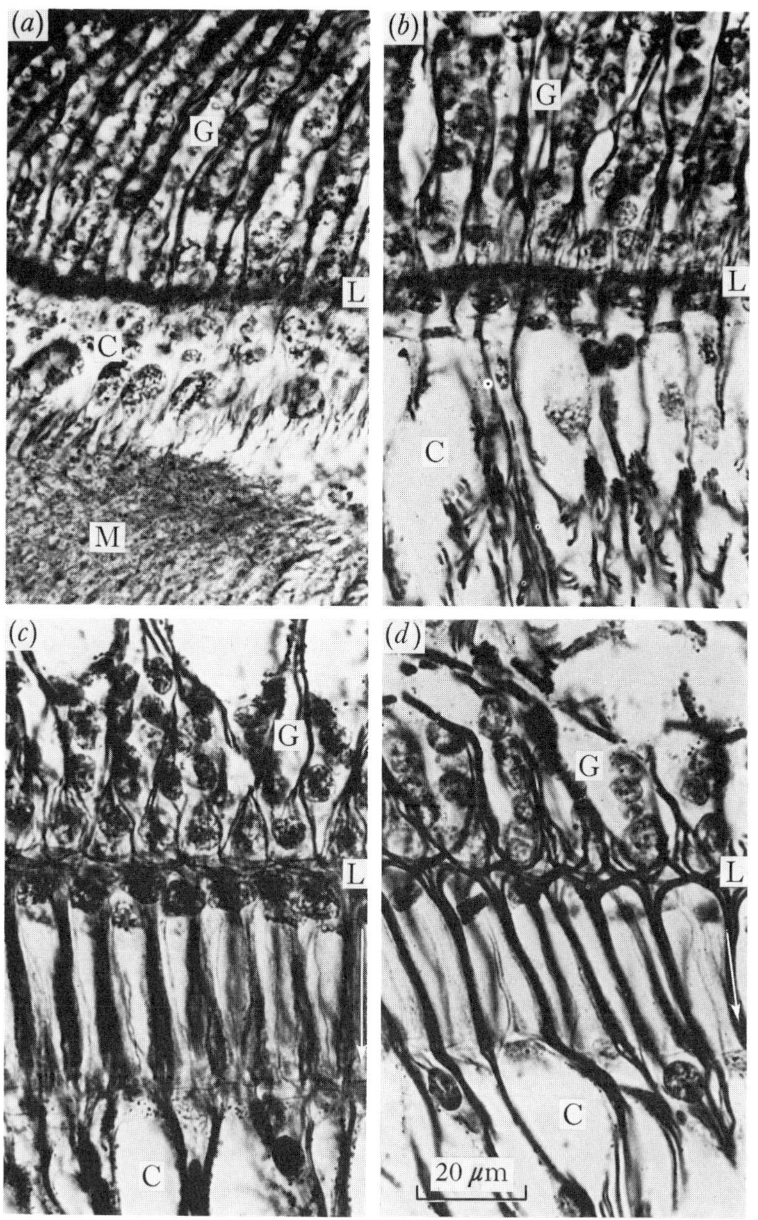
(a)
G
L
C
M
(b)
G
L
C
(c)
G
L
C
(d)
G
L
C
20 μm

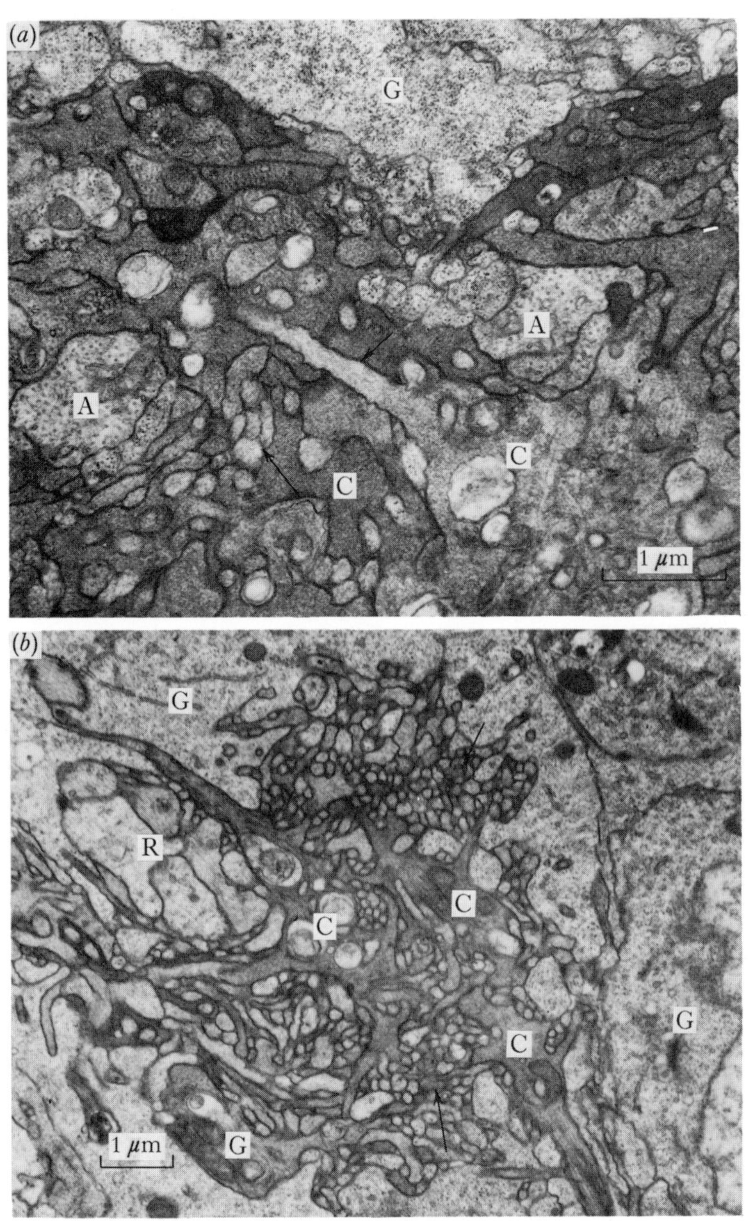
(a)
G
A
A
C
C
1 μm
(b)
G
R
C
C
G
C
1 μm
G

Development of the lamina neuropile*

Although earlier evidence (Power, 1943) suggested that the lamina arose from the eye disc, direct developmental evidence (Gieryng, 1965; Shatoury, 1956, 1963) shows that the lamina in common with the other optic neuropiles develops from the brain.

Innervation from the optic stalk. Ommatidial axon bundles arrive at the lamina cortex from the eye disc in predominately a postero-anterior sequence. The growth sequences of axon bundles are explained in the caption to Fig. 2 and their organisation probably results directly in the ordered retinotopic projection of bundles between retina and lamina. Ommatidial bundles of retinula axons grow to the edge of the lamina cell cortex after entering the larval hemisphere through the optic stalk. Growing over the region already innervated by retinula axons until they have spread over the first vacant cell group they then plunge centrally, on the side of the monopolar cell group furthest from the equator. As a result of the growth pattern between retinula and monopolar axons the position of the retinula axon bundle reverses, and twisting 180°, finishes on the side of the monopolar cell group nearest the equator (Hanson, Jiang & Lee, 1972). Axons grow through the optic stalk without changing positions within their bundles, i.e. without the bundle braiding, so that their original retinal orientations are preserved at the lamina. It is not clear whether the long retinula axons of each bundle grow before the short retinula axons as in *Pieris* (Sánchez, 1919*a*) and *Agrion* (Richard & Gaudin, 1960) (see p. 61).

* The outline of lamina development presented in this section is from unpublished electron microscopic observations by Hanson, Jiang & Lee on *Drosophila* and by Meinertzhagen on *Calliphora* and *Lucilia*. The observations agree in all points of overlap and to avoid repetition authors are quoted only when they have made the sole observation.

Plate 2. Electron micrographs of the lamina plexus in (*a*) mid third instar and (*b*) late third instar larvae of *Calliphora stygia*. The region of the plexus in (*a*) is from near the growing edge and shows growth cones C and filopods (arrows), during the initial events of filopod invagination, with two rows of monopolar axon bundles A. At this stage the growth cones have regular outlines while at a later stage (*b*) they are drawn out in a number of directions. The plane of section in (*b*) grazes the lamina plexus superficially and shows the growth cones of one retinula axon bundle with a neighbouring axon bundle R growing over the surface of a monopolar ganglion cell body G.

The lamina plexus and the growth cones of short retinula axons. The characteristic feature of the adult lamina, the columnar synaptic neuropile of the cartridges, does not make its appearance until half way through pupal life (e.g. 4–5 day pupa, in *Calliphora*, Plate 1) (Gieryng, 1965).

Throughout larval and early pupal life the lamina is represented by a thin sheet, the lamina plexus. The plexus contains the growing terminals of the short retinula axons penetrated at intervals by bundles of lamina monopolar axons and long retinula axons which enter the external chiasma and grow down to the medulla.

On entering the plexus the short retinula axons form large growth cones* (1–5 μm diam.) which swell to contact each other and are limited on the internal surface of the plexus by a layer of cells of several types, including glial cells and amacrine neurons. The growth cones give rise to a large number of filopods which invaginate into neighbouring growth cones often for long distances and in a highly redundant manner. The initial process of invagination is known mainly from observations on the larvae of *Calliphora* (Plate 2) (Meinertzhagen, unpublished). It appears to be non-selective because filopods invaginate not only into growth cones of other axon bundles but also of their own bundle and even on occasion into another part of their own growth cone. They also invaginate superficially into the cytoplasm of the underlying cells but do not however penetrate the monopolar and long retinula axon bundles. The very first filopod to invaginate into one growth cone from another seems to be in the centre of the area of membrane contact between the two and is frequently followed by a filopod being formed next to it from the second to the first. The filopods are slender (0.15 μm diam.) and straight, often travel along each other, and frequently grow coaxially into other filopods thereby gaining access directly to subadjacent growth cones. As they grow, the filopods soon become orientated preferentially along a number of directions, probably as a function of the angle at which growth cones come initially into contact in a regularly spaced array.

Soon after these initial events the growth cones become so completely invaginated that their extents become difficult to assess

* It is assumed that the growth cones described here correspond to those seen in Golgi profiles in the developing optic lobe of *Pieris* (Sánchez, 1919*a*, *b*) and the filopods to those first described in the rabbit dorsal root neuroblast (Tennyson, 1970).

and they become connected with a large number of other growth cones. They elongate progressively and spread across the lamina but the total area invaded by an axon of identified retinula cell class is not yet known. Finally a group of growth cones forms around a bundle of monopolar axons separating these from their accompanying long retinula axons which come to occupy a satellite position. The later process of growth cone elongation is known mainly from observations on the pupae of *Drosophila* (Hanson, Jiang & Lee, 1972). Early in development the growth cones of one ommatidial axon bundle are connected together by junctions which form a circle surrounding a group of monopolar axons, while later (48 h after pupation in *Drosophila*) when the growth cones have spread across the plexus these original junctions have disappeared and new ones formed around the monopolar axon bundle between the convergent growth cones which will contribute to the formation of that cartridge (Hanson, Jiang & Lee, 1972).

Individually these filopods probably have no permanent significance in the establishment of the adult connections but their total pattern is responsible for the spreading of the growth cone across the lamina so that it occupies a range of positions in the plexus at different times. Subsequent development involves selective regression of all but one connection as the plexus condenses to form the final pattern of adult connections. This process of condensation results in the exchange of short retinula axon growth cones between neighbouring ommatidial axon bundles so that each growth cone becomes associated with only one set of lamina monopolar axons, some distance from its original axon bundle. These developmental events occur mainly before synapse formation. For example, in *Drosophila*, growth cones have reached their cartridge sites by 48 h after pupation, but the formation of recognisable synapses between retinula and monopolar axons does not occur until 72 h after pupation (Hanson, Jiang & Lee, 1972). The cartridges eventually form by centripetal growth of the ring of growth cones down the monopolar axons accompanied by an inward shift of the amacrine and lamina epithelial cell nuclei (Plate 1).

The overall sequence of events has more than a passing similarity to those found in a number of vertebrate systems (e.g. Morest, 1969) including the maturation of retinotectal connections in *Xenopus* (Gaze, Chung & Keating, 1972), i.e. the development of individual nerve connections by progressive and selective restriction of neuronal fields. The systems differ however in that final connections are made in the fly before the establishment of synaptic contacts.

Development of the medulla neuropile

The medulla neuropile develops around the 'ocellar' nerve which runs along the equator of its external surface (Fig. 3).

The first sign of the medulla neuropile is the early differentiation of the medulla tangential neurons which grow out axons to meet the first perpendicular neurons of the medulla. Subsequently the horizontal growth of tangential cell processes continues to form an ever-expanding weft as more perpendicular neurons differentiate and add new threads to the warp of the neuropile. In spite of their early production and differentiation, the development of these horizontal processes continues right into the late pupa (Strausfeld & Blest, 1970).

Formation of chiasmata

Axon bundles leaving the internal face of either lamina or medulla enter a wedge-shaped tunnel which is bounded by ganglion cell groups on either face (Fig. 3). The tunnel itself contains groups of cells scattered regularly throughout it, some very large which probably give rise to the glial cells of the external limb of the two chiasmata. The youngest axon bundles grow along the interface of the tunnel wall formed between axon groups which have already grown and an adjacent ganglion cell mass (medulla cortex, external chiasma; lobula cortex, internal chiasma). Ganglion cells bordering this interface and in close proximity to the new axon bundles begin differentiation and send their axons out to join the new wave of chiasmal fibres. The axon bundles therefore initially grow at close quarters between the neuropiles and subsequently elongate as more recent neurons are added in the sequence.

Within this sequence the ordering of the chiasmal strata results from the pattern of paths taken by the growing axons, but these are clear only in the case of the external chiasma (Fig. 4). Further information is required to clarify the following points in the sequence of axon growth within each external chiasmal bundle. It is not known whether the pair of long retinula axons grow synchronously or sequentially, although a simple explanation of

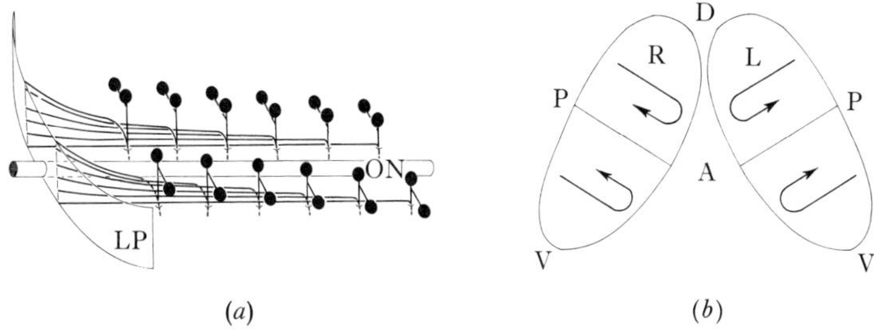

Fig. 4. Diagrams to explain how the organisation of the strata of the external optic lobe chiasma is produced by the patterns of axon growth within the growth sequence across the eye.

(*a*) The growth pattern of two chiasmal strata on either side of the equator. The first axon bundles growing from the equatorial region of the posterior margin of the lamina plexus LP follow on either side of the 'ocellar' nerve ON. Axon bundles added subsequently from the next horizontal rows take paths parallel to the first while those from the same horizontal rows run along the pre-existing bundles. There is no confusion between horizontal bundle rows because at the growing edge of the lamina the separation between them is far greater than that between vertical rows (which are nearly confluent). The twist within each stratum is produced by the paths followed by these subsequent bundles. Axon bundles recruited anteriorly in each horizontal row invert their sequence in the row by deflecting internally (i.e. equatorially) on either side of the equator, before penetrating the medulla neuropile together with the axons of transmedullary and medulla centrifugal cells. At the internal face of the medulla cortex the axons of these cells are stained by reduced silver and their concentration gives a neuropilar appearance to this region.

(*b*) The direction of twist of the strata of the adult external chiasma. In passing between lamina and medulla the chiasmal axons invert their horizontal sequences by a clockwise twist on one side of the equator and an anti-clockwise twist on the other, with opposite directions of twist in the other eye, as indicated (Braitenberg, 1970).

the different depths in the medulla neuropile at which their terminals are located may lie in one reaching the medulla first, when the neuropile is less complex. Retinula axons going to lamina and medulla could transfer directly information on the field size of centripetal innervation while the chiasma could be formed initially by the long retinula axons which would serve as guidelines for the subsequent growth of axons as Sánchez describes (1919*b*). Any growth sequence between long retinula and monopolar axons must however be quite rapid because even in chiasmal bundles no more than a few hours old two long retinula and at least three monopolar axons are present. Probably medulla centrifugal cell processes grow after the initial lamina–medulla connection has been established by the centripetal neurons, because retinula and monopolar axons are the first to be found in the lamina plexus. This growth sequence suggests that the centrifugal cell axons grow back along their chiasmal axon bundle.

The significance of chiasmata

It will be obvious from the foregoing account that the external chiasma develops as a simple and necessary consequence of the pattern of normal growth (Pflugfelder, 1937). Probably it has no greater significance than this. There are however two qualifications to introduce to this generalisation.

Firstly there are persistent reports of perpendicular neurons with uncrossed axons running in the external chiasma, stemming from the important early observations by Cajal & Sánchez (1915).* Almost certainly these are either erroneous observations or aberrant fibres (see Strausfeld, 1971*b*). At least some fibres, from lamina tangential neurons, do however appear to decussate in an atypical fashion (Strausfeld, 1970) and their part in the lamina–medulla pattern of axon growth must be substantially different to that of the perpendicular neurons.

The second possible qualification concerns small differences in the pattern of axon growth reported in *Pieris* involving interweaving between the axons as they invert their horizontal sequence in the external chiasma (Fig. 8*b*). (Sánchez, 1920).

If chiasmata are simply a consequence of the normal growth pattern one might well ask which is the fundamental process, the sequential growth pattern across the eye or the formation of chiasmata.

Some interesting comparisons may be made with the development of crustacean optic lobes (Elofsson & Dahl, 1970; Parker, 1897) which provide a partial answer to this question. The pattern of axon growth between lamina and medulla externa for the perpendicular neurons of non-Malacostraca is shown in Fig. 5(*b*) and of Malacostraca is shown in Fig. 5(*c*). In both groups growth occurs as a sequence across the eye but only Malacostraca possess chiasmata. This essential difference between the two types arises from the direction from which the monopolar axons of the lamina

* See their Fig. 15 for bee and Plate 2 for fly.

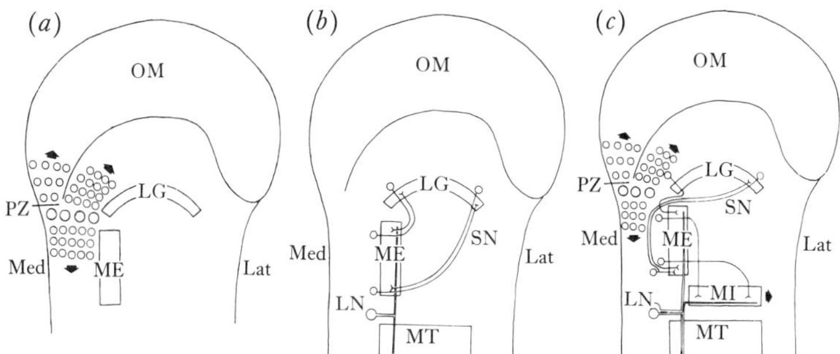

Fig. 5. The development of the crustacean compound eye and optic lobe showing schematically, from the dorsal surface, the patterns of anlage proliferation and fibre growth (from Elofsson & Dahl, 1970).

(*a*) A medially situated semicircular proliferation zone PZ produces the ommatidial cells of the retina OM and the cartridge cells of the lamina ganglionaris LG and medulla externa ME. The initial orientation of lamina ganglionaris and medulla externa is essentially orthogonal.

(*b*) Arrangement of the fibre connections in an adult non-malacostracan. The most central of the optic neuropiles is the medulla terminalis MT. The arrangement of the 'long' neurons LN, the tangential cells of the medulla, and the 'short' neurons SN, the perpendicular cells of the lamina and medulla, is shown. The fibre growth pattern of the perpendicular neurons occurs between the two neuropile surfaces and produces an uncrossed projection. In some groups the medulla externa rotates in a counterclockwise direction to occupy a near medio-lateral orientation.

(*c*) Post-embryonic arrangement of fibre connections in a malacostracan. In these higher Crustacea the medulla externa is represented by two intermediate neuropiles, medulla externa and interna MI. The fibre growth patterns of the perpendicular neurons between lamina and medulla externa occur between the neuropile surface of the lamina and the cell body surface of the medulla externa and give rise to a chiasma between the two. This is most easily seen in the adult because the medulla externa undergoes a clockwise rotation in subsequent development. Lat, lateral; Med, medial side of the eyestalk.

first approach the neuropile of the medulla externa. In non-Malacostraca they approach the medulla externa from the side of the neuropile, in Malacostraca from the side of the cell cortex (Elofsson & Dahl, 1970) thus forming the chiasmata in essentially the same way as is found in insects.

In other words, sequential growth is fundamental to all compound eyes but the formation of chiasmata is not a necessary consequence of it.

The formation of nerve connections in the optic lobe

The formation of nerve connections is one of the most stimulating and perhaps the most challenging aspect of optic lobe development. The most important principle yet to emerge is to what extent neurons find termination positions which are specified or labelled for them or alternatively, to what extent they form specific connections as a result of the restraints imposed on them by the temporal and spatial sequences of the growth process. Distinction between the two alternatives is possible only by experimental intervention, one approach being transplantation and regeneration experiments in which neurons are presented with totally new situations in which to express their growth tendencies. In order however to propose explicit mechanisms it is necessary to have an exact knowledge of the behaviour of individual neurons in normal growth.

Evidence from regeneration: the macroscopic projection of retina upon lamina

If the retinotopic projection between retina and the central optic neuropiles arises *de novo* as a result of the ordered axon growth sequences between the different layers, do separate regions of these central neuropiles become differently marked (labelled or specified) as a result of their innervation, in some way which could be used by regenerating axons to find their correct position during regrowth?

This type of question has received extensive attention in the amphibian retinotectal projection from a number of workers (Gaze, 1970); it contains conceptual and semantic difficulties which are well recognised (Gaze *et al.* 1972). Some preliminary evidence is available in insects. Experiments on locusts (Horridge, 1968) at first sight suggest that the lamina is not differently labelled as a result of having received innervation from the retina, at least between dorsal and ventral areas. Second instar nymphs of *Schistocerca* in which the retina was rotated through 180° and allowed to regenerate, showed normal visuomotor behaviour in the adult. In assessing these experiments however it is necessary to bear in mind that in the second instar nymph less than half of the adult complement of ommatidia is present (Bernard, 1937) and that the ommatidia formed subsequently are free to grow their axons *de*

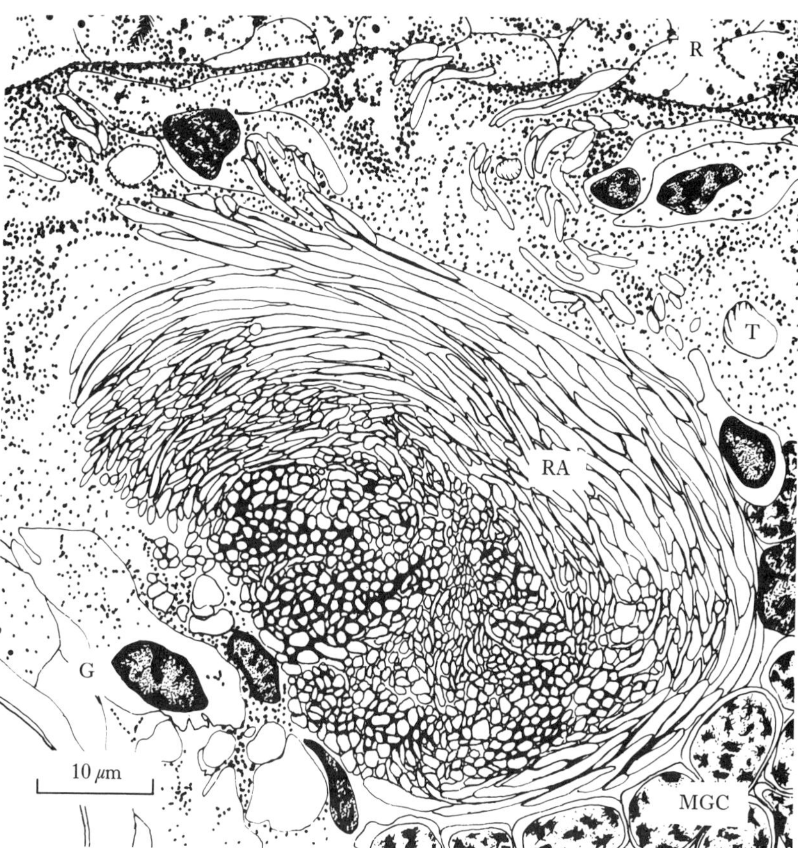

Fig. 6. Diagram of the retinula axons RA in an adult locust *Chortoicetes* beneath an area of retina R which had been cut through in the third instar nymph. The regenerating ommatidia produce axons which, probably because they grow back synchronously or in a temporally disordered sequence to the lamina, become entwined into large spherical whorls within which the axons wind around each other in a manner reminiscent of the continuous circles of lost regenerating sensory axons in the epidermis of *Rhodnius* (Wigglesworth, 1953). The axon whorls abutt against the layer of lamina monopolar ganglion cells MGC and are surrounded by glial cells G and trachea T. Apparently few axons leave this whorl and enter the lamina neuropile. (Preparation courtesy of Dr C. M. Bate).

novo. Since in the adult their number exceeds that of the old regenerated ommatidia they would be expected to dominate vision (P. M. J. Shelton, personal communication). No conclusions can therefore be drawn from eye rotation experiments which fail to distinguish between old (i.e. pre-operative) and new (post-operative) ommatidia.

In experiments in which a cut was made across half the vertical height of the retinae of second instar locust nymphs, most animals showed behavioural recovery which is uniform throughout the entire old part of the eye (J. Kien, personal communication) suggesting that retinula axons must be regenerating at points right across this region. If these same experiments are carried out on third instar locusts there is only some recovery of visuomotor function in the old ommatidia, which is confined to small areas surrounded by blind facets (Fig. 6) (J. Kien, personal communication). In these areas, as in those from animals operated on in the second instar, visuomotor efficiency is greatly reduced indicating that the normal pattern of connections is not completely regenerated. On the other hand the new ommatidia continue to grow their axons down the edge of those old retinula axons which remain uninterrupted (C. M. Bate & J. Kien, personal communication).

Where the eye of a third instar locust is sliced off at a level sufficiently distal to leave the growing anterior margin *in situ* and rotated 180°, axons of new ommatidia can be seen spanning the retina–lamina projection in apparent ordered array, while very few old ommatidia regenerate axons (P. M. J. Shelton, personal communication). The old lamina remains unoccupied, the new axons occupying only the most recent (anterior) lamina, which directly underlies them and which they possibly reached by simply following the last grown of the pre-operative retinula axons before these degenerated. On the other hand recovery amongst the old ommatidia in animals of this age is apparently limited by the poor regeneration of retinula cells, which possibly have either insufficient time or lose the ability to regrow their connections with the lamina.

In summary, the ordered retina–lamina projection found in the locust (Meinertzhagen, 1971) probably cannot arise under conditions of synchronous axon regrowth, i.e. without the sequential growth of ommatidial axon bundles that occurs in normal development. If behavioural tests are to be used to analyse regenerated connections this would seem to place a limitation on the usefulness of compound eye regeneration as a precise test of the specification of the lamina, resulting from prior innervation. It does not of course preclude its use to test for specification at a more regional level but experiments are yet to be reported in which the analysis of regenerated connections gives evidence of such specification.

Evidence from growth studies

Interneuropilar and intraneuropilar fibre growth. The axon growth patterns of the perpendicular neurons of the fly optic lobe (p. 77) reveal two phases of growth which, though intimately related and continuous with each other, have different characteristics and perhaps require different mechanisms. The first type of growth is interneuropilar and occurs in an ordered temporal sequence by which the centripetal growth of axon bundles is converted to an ordered spatial sequence of parallel elements between two adjacent cell populations (retina–lamina; lamina–medulla; medulla–lobula). This gives way to the second, intraneuropilar type of growth, which results in the establishment of specific connections in the neuropile. Apart from their temporal separation, the two types of growth, which correspond roughly to the distinction between axonal and dendritic growth of vertebrate neurons (Jacobson, 1970; Morest, 1969), are distinguishable by several criteria.

Interneuropilar growth is rapid and probably depends on some form of mechanical or contact guidance, because it always occurs at the interface between a population of cells and groups of axons already connected. The axon bundles probably retain their spatial order primarily because they grow out singly in any one region. The smooth growth cones of retinula axons in *Pieris* (Fig. 8*f*, *g*), growing between neuropiles (Sánchez, 1919*a*), indicate that filopod activity is minimal and axon growth presumably largely non-exploratory. The distances over which interneuropilar growth occurs are altered by subsequent development, as in a variety of vertebrate peripheral axons. In the two chiasmata the axonal connections are established at close range and later elongate in a manner similar to the formation of the interganglionic pathways of insect ventral cord connectives, e.g. in *Pieris* (Eastham, 1930). In the retina–lamina projection the axon connections grow over a long distance but along a preformed path initially provided by the 'ocellar' axon bundle, or, in Lepidoptera, the stemmatal nerve. The lengths of these axons in the adult are attained largely by the shortening of the optic stalk axons, simultaneously drawing the lamina towards the retina and elongating the two chiasmata (Shatoury, 1956).

Intraneuropilar growth is known in detail from EM observations of the fly lamina plexus during the formation of specific retinula–lamina connections (Hanson, Jiang & Lee, 1972; Meinertzhagen, unpublished). In the lamina of the fly this type of growth consists of a temporary and promiscuous proliferation of lateral filopods from the enlarged growth cone of the axon tip, with the subsequent condensation of these structures to form a single final axon connection by selective regression of incorrect connections. Intraneuropilar growth is therefore a long-term intercalary growth process.

Observations by Sánchez (1919*a*) provide support for a similar type of growth in the retinula terminals in the lamina of *Pieris* (Fig. 8*c*). Furthermore they indicate that, as in *Calliphora*, the short retinula terminals of adult *Pieris* are secondarily simplified during development, i.e. they have greatly reduced numbers of lateral processes compared with their immature forms (Figs. 7 and 8). This extreme simplification is not found in all adult insects, e.g. bee (Cajal & Sánchez, 1915) and locust (Horridge, 1968), where some lateral processes remain. Probably the intraneuropilar growth process is essentially similar in the short retinula terminals of all insects, even though in adults of some species such as the fly and *Pieris* the terminals are structurally little more than the prolongation of their axons and even though the axon connections for fused- (*Pieris*) and open-rhabdome (fly) ommatidia are different (Meinertzhagen, 1971). Electron microscopic observations of the developing central optic neuropiles and also of the immature neuropile of the dorsal ocelli in *Drosophila* reveal similar arrangements of growth cones and filopods (Hanson, Jiang & Lee, 1972) so that a basic understanding of the behaviour of growth cones in any one region is certain to have widespread significance in the analysis of neuropile development.

Growth cone evolution. Until recently the only detailed accounts of the morphology of growth cones in any developing invertebrate

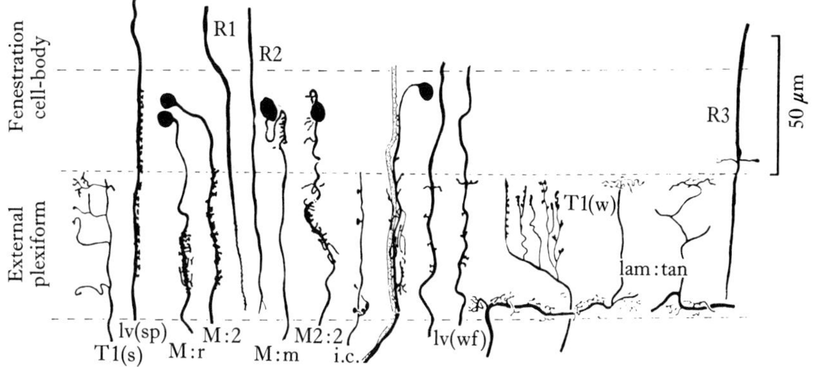

Fig. 7. Diagram of the elements of the adult lamina of *Pieris brassicae* for comparison with the immature counterparts shown in Fig. 8. Three types of simple short retinula terminals R1–R3 are present together with three types of long retinula axon which terminate in the medulla. The long retinula axons have different types of processes in the lamina, either spiny lv(sp), wide-field lv(wf) or smooth (stippled). Five forms of monopolar cell are also present (From Strausfeld & Blest, 1970. Reproduced by permission of the Council of the Royal Society.).

nervous system were Sánchez' monumental yet largely unquoted Golgi studies on optic lobe development in *Pieris* (Sánchez, 1916, 1918, 1919*a*, *b*). Although this author uses at least 14 adjectives to describe the shape of these organelles, there are two basic types: smooth, assagai-shaped ones at the tips of freely growing axons and the large, filopod-covered, expanded growth cones of the neuropile.

It is possible to distinguish three phases in the development of the expanded growth cones, most clearly seen in the retinula and lamina monopolar neurons (Fig. 8).

1. A *tumescent* phase which results in the axon expanding to form a smooth growth cone. Usually this occurs at one point and later extends axially along the axon. The expanded growth cone probably forms from the impedance, within the neuropile, of the much smaller growth cone found at the axon tip during inter-neuropilar growth, since retinula axons often form transient swellings if temporarily obstructed in their passage through the basement membrane of the retina. Formation of expanded growth cones by premature impedance may be the mechanism by which a hernia of the medulla (Strausfeld, 1971*a*) was formed. In the long retinula axons however growth cones are found not only at

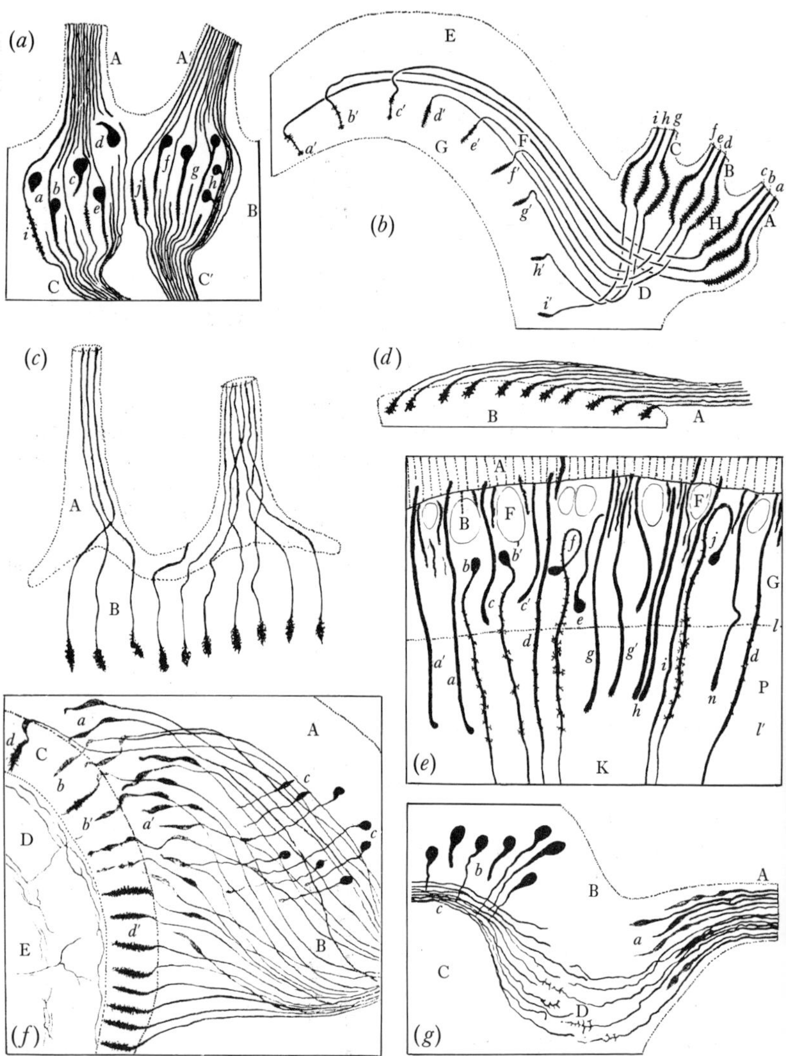

Fig. 8. Diagrams of developing optic lobe neurons of *Pieris* pupae, from Golgi impregnations.

(*a*) Young lamina with monopolar cells either commencing axon outgrowth (*a*, *c*, *d*) or just completed axon outgrowth (*b*, *e*, *f*–*h*). Profiles *i* and *j* are the lamina arborisations of long retinula axons. *A*, *A′* columellae of retinula axons; B lamina; C, C′ axon bundles of the external chiasma.

(*b*) Scheme for the formation of the external chiasma in *Pieris* according to Sánchez. Long retinula axons *a*–*i* already have well-formed growth cones in the their termination in the medulla (Fig. 8*d*) but also some time later in the developing lamina, after their axons have grown through this region (Fig. 8*b*). In this case the formation of a growth cone appears to be induced by the region surrounding it, perhaps by the presence of other growth cones which have formed with the invasion of the short retinula axons. Possibly the formation of a growth cone requires some change in the property of the axonal membrane.

2. Shortly after the growth cone swells the first *filopod growth* occurs, presumably corresponding to the invagination into neighbouring growth cones. Filopods continue to be produced during the last phase of growth cone expansion and for some time afterwards, until the whole surface becomes covered evenly. They appear rather constant in length for each class of neuron.

3. The final phase consists of the *detumescence* of the growth cone and the gradual loss of filopods to leave the patterns of lateral processes characteristic of the adult termination, as if those

---

lamina. The axons decussate in the chiasma D and penetrate between ganglion cell layer E and medulla neuropile G before forming the corresponding terminals *a′–i′*. The axons arrive through three columellae, first through A, then B, then C and the sequence of their ages is illustrated by the form of their medulla terminals, *a′* being oldest, *i′* youngest. In inverting their horizontal sequence the axons interweave not only with those of their own columellae but also with those of older columellae, so that judging from this figure there is no coherent twist in a chiasmal stratum but a number of local intercrossings.

(*c*) Short retinula axons in a young pupa entering the lamina B through two columellae A and forming expanded growth cones.

(*d*) Bundle of long retinula axons A at their entrance to the medulla neuropile, showing the early appearance of their expanded growth cones and filopods.

(*e*) Part of the lamina of an older pupa after the almost complete maturation of the processes of retinula and monopolar axons. Some cell types are illustrated by two congeners. *a*, *c*, *h*, *n*, are short retinula axons; *d*, *i* long retinula axons and *b*, *e*, *f*, *j*, monopolar axons. A retina, F, F′ fenestration zone, B basement membrane, G ganglion cell zone, P lamina neuropile, K external chiasma.

(*f*) Outer part of the medulla showing the form of growth cones during free growth in the external chiasma of long retinula axons B or axons of trans-medullary cells *c* from the medulla ganglion cell layer A. C, D, E are strata of the medulla neuropile; *b*, *d* are the retinula terminals in the most superficial stratum.

(*g*) The base of the first retinal columella A from an early pupa, showing freely growing retinula axons *a* entering the immature lamina D and others which have already grown to the medulla C. ((*a*), (*c*)–(*g*) modified from Sánchez, 1919*a*; (*b*) modified from Sánchez, 1919*b*).

connections remaining had a more effective adhesive affinity for their partners than those that regressed. Frequently whole regions of the growth cone contract to the diameter of the axon and lose their filopods entirely, so that the final termination is stratified in different vertical layers of the neuropile.

Retinula cell connections in the fly lamina

The connections of individual retinula cells are in general specific and relate to the position of the cell within the ommatidium (Fig. 9*a*) (Horridge & Meinertzhagen, 1970). This leads to the problem of how the axon classes differ in order to form their different connections, i.e. what is their degree of specification. For example, if the differences in their final connections could be explained solely in terms of their spatial arrangement when they first arrived at the lamina or their intraommatidial sequence of arrival, we would not need to infer that they were differently specified from each other. Of course an explanation would still be required of how the axons came to occupy that spatial arrangement in the lamina or how they came to arrive in that temporal sequence.

The distribution of retinula axons within a cartridge. In all but the equatorial cartridge rows, the crown of short retinula terminals in one lamina cartridge is in a sequence 1–6, with each terminal coming from a different ommatidium (Fig. 9*b*). This ordered sequence of retinula terminals around the circumference of a cartridge possibly results from the coincidence of three circumstances in the growth process. First, it seems that the connections of the adult axons simply terminate in the position at which they first encounter a suitable cartridge, whether this is the correct cartridge or not (Meinertzhagen, 1972), whether they have crossed the equator or not and regardless of axon class (Horridge & Meinertzhagen, 1970). Second, elongation of the growth cones is in the straight lines laid down by the exploratory filopods. Third, the pattern of ommatidia, each contributing a different short retinula axon to a particular cartridge, itself forms a spatial sequence around that cartridge (Fig. 9*b*). Most likely the distribution of terminals within a cartridge merely reflects this pattern of convergent retinula axons.

Occasional inconsistencies in positional sequence of terminals within a cartridge in those examples so far studied can all be attributed to transpositions of retinula terminals 3 and 4, usually between themselves and most frequently in those that cross the equator (Horridge & Meinertzhagen, 1970; Meinertzhagen, 1972). Thus the occurrence of these transpositions would be explained by the fact that the angular sector occupied by ommatidia which will contribute axons 3 and 4 to a cartridge is smaller than the sector occupied by ommatidia contributing any other two neighbouring terminals (Fig. 9*b*).

The distribution of retinula axons to different cartridges. Schoeller (1964) found that retinula axons grow into long disorganised bundles after *in situ* optic stalk section in *Sarcophaga*. Apparently the formation of short retinula axon connections is not intrinsic to the retina but depends upon interaction with the array of lamina monopolar cells. Nevertheless each retinula axon appears to behave autonomously in forming its connection (Meinertzhagen, 1972), although it probably uses the growth cones of other axons to do so.

Some evidence on the importance of the spatial arrangement of axons when they first arrive at the lamina comes from studying patterns of abnormal connections of retinula axons, the general principle being that the nature of growth rules is revealed by the category of erroneous connection that they permit. This approach is admittedly imprecise and firm conclusions will only be reached when a full description of the developmental events becomes available.

What happens to retinula axon connections when spatial order in the lamina is disrupted? A number of mutants have been observed for example, in *Drosophila*, which have aberrations in both retinula cell pattern and lamina neuropile organisation (Hanson, Jiang & Lee, 1972). One of these has almost randomised retinula axon connections, from which it was concluded that retinula axons are capable of establishing a wide variety of erroneous connections, i.e. they have no mechanism which restricts them from doing so. Greater consistency in a number of erroneous connections was seen when the degree of disorder was more subtle. In *Calliphora*, 16 axons associated with a dislocation in the retinal equator were found to terminate in the wrong lamina cartridge (Fig. 9*c*) (Meinertzhagen, 1972). The frequency of error was highest in those axons which normally travel furthest laterally

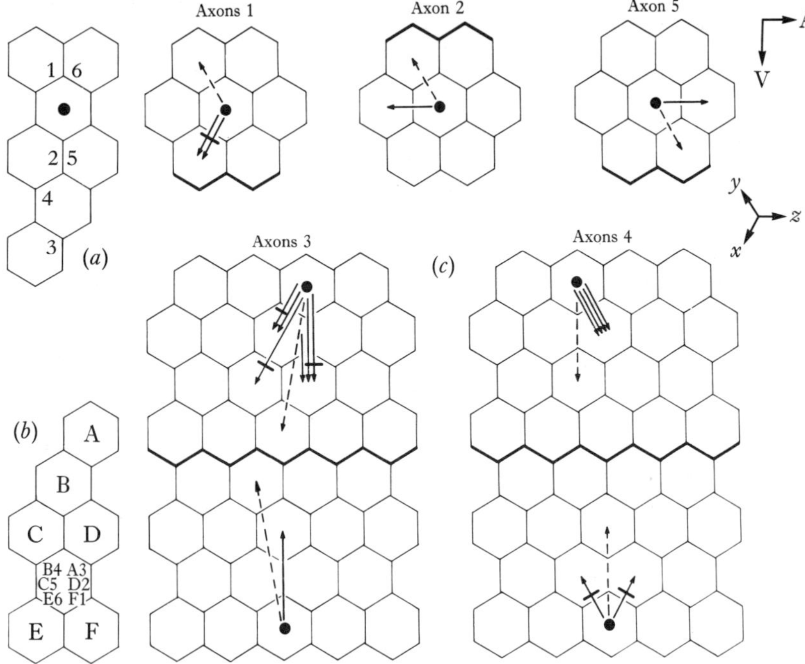

Fig. 9. The organisation of the first visual projection of the adult fly. Each ommatidium of the retina of a fly contains eight retinula cells, each retinula cell bearing along its inner longitudinal margin a separate photoreceptive structure, the rhabdomere. In transverse section the rhabdomeres of one ommatidium form a distinctive asymmetrical pattern within which each cell may be uniquely numbered. The retinal array of these rhabdomere patterns has a sharp discontinuity about a horizontal equator so that the pattern of an ommatidium in the ventral half is the mirror image of that in the dorsal (Dietrich, 1909). The retinula cells 1–6 of a single ommatidium have opposite rotational sequences in the two halves. Two central cells 7 and 8 lie within the ring of retinula cells 1–6 and give rise to a pair of long visual fibres which proceed through the lamina to the second optic neuropile, the medulla. The surrounding cells 1–6 have short retinula axons which terminate in the lamina cartridges, the serially repeated structures of this neuropile. Six short retinula axons from retinula cells 1–6 of six different ommatidia converge upon a single cartridge of the lamina (Braitenberg, 1967; Trujillo-Cenóz & Melamed, 1966) where they surround a group of five monopolar neurons L1–L5 (Strausfeld, 1971*a*).

(*a*) The axon bundle of a dorsal ommatidium is represented by a solid circle at the centre of a pattern of six lamina cartridges each of which is represented as a hexagon and receives one short retinula axon terminal. The pattern of projection of the six axons is predictable and there is a set spatial relationship for each axon class between the ommatidium of origin and cartridge of termination.

in the lamina, i.e. in classes 3 and 4, which travel in approximately the right direction but stop short of the correct distance, terminating preferentially in a cartridge next to but nearer than their predicted one. On the other hand, Strausfeld (1971*a*) illustrates axons which travel unusually long distances in the lamina. The simplest interpretation is that cell recognition of specific monopolar axons is not the method by which axons reach their cartridges. In other words, monopolar axons are probably not labelled uniquely according to their position in the lamina, although they must be distinguishable as a class by the advancing retinula axon growth cones, from amongst the other cellular elements in the lamina plexus. It seems more likely that axons simply terminate in the cartridge in which their growth cone is most prominently represented, when the time comes for cartridge formation, but are normally precluded from staying with their original monopolar axon bundle. On occasion this may result in an axon being represented in two cartridges (Strausfeld, 1971*a*). The final confirmation of the group of convergent growth cones which will contribute to one cartridge is manifest by the formation of a circle of junctions between them (Hanson, Jiang & Lee, 1972). Most obscure at the moment is the mechanism by which growth cones become interchanged between neighbouring axon bundles. Perhaps filopod competition between neighbouring ommatidial growth

---

Axons 2–5 spread towards the equator and axons 1 and 6 away from it in either half of the eye (Braitenberg, 1967).

(*b*) The pattern of convergent retinula terminals for a cartridge in the dorsal half of the lamina. The rotational sequence of terminals is 1 to 6 and is anticlockwise (clockwise for ventral cartridges). The retinula terminals are derived from a pattern of ommatidia A–F surrounding the cartridge, one from each. The pattern is a mirror-image in the ventral eye half.

(*c*) The projection patterns of 17 axons which terminated in incorrect cartridges in a total plot of short retinula axons from a region of a right eye containing a dislocation in the retinal equator. Five separate patterns are shown for axons 1, 2, 3, 4 and 5, plotted relative to a common origin. The point of origin for each axon is represented by a solid circle in each plan and the solid arrows represent the paths of axons to their cartridge. Broken arrows represent the predicted pathway for each axon class. Projection patterns above the heavy horizontal lines are from the dorsal half of the eye and those below are from the ventral. Arrows crossed by short heavy lines represent the paths of axons that cross the equator to reach their cartridges. Axes, x, y, z are defined by the orientation of retinal facet rows. A, anterior; V, ventral (From Meinertzhagen, 1972).

cones ensures that each axon class diverges in its own unique direction.

The evidence suggests that the orderly connection of retinula axons is a reflection of the orientation and spatiotemporal order with which the axons become situated in the lamina. This order probably originates from the developing retina through the cellular arrangement and differentiation within the ommatidia, and is communicated to the lamina by the patterns of retinula axon growth. Especially significant is the way in which the orientation of the axons within their bundles is maintained through the optic stalk and how the orientation is twisted through 180° as a result of passing through the lamina cortex as Hanson describes. In doing so the monopolar axons always come to be situated between retinula axons 1 and 6. The twisting must reflect the arrangement of retinula cells within the ommatidium, because both are in opposite senses in dorsal and ventral halves of the eye. Perhaps the twisting arises by a sequence of axon outgrowth determined by the retinula cell arrangement but, whatever the mechanism, in the fly the development processes are refined to a level of remarkable order and regularity, as with so many features of its visual system.

Conclusions

It is clearly too early to summarise the evidence for specification in retinula axon connections. The evidence seems to exclude specific cell recognition as a factor but to implicate spatial order, which with reference to the 'conveyer belt' development characteristic of insect optic lobe, in turn implies that temporal order must be important.

The principles beginning to emerge from the study of the optic lobes of flies almost certainly apply but probably with less regularity to the morphologically less highly ordered optic lobes of other insects. The success of studies of this sort on optic neuropiles will however ultimately lie in the degree to which the conclusions may be applied to the genesis of those neuropiles in which order is not apparent.

Summary

1. A brief review is given of the main events in the development of compound eye and optic lobe in insects, with particular refer-

ence to the formation of nerve connections in the fly optic lobe.
2. In general the developmental events occur in precisely controlled temporal sequences which generate spatial order in the optic lobe neuropiles. This principle operates in two ways, first in the proliferation of retina and optic lobe, then the subsequent differentiation of both.
3. Almost certainly the retina develops autonomously of the optic lobe in all insects. On the other hand differentiation of the cells of the optic lobe is dependent upon retinal innervation, even in some cases for the stimulus for axon outgrowth.
4. Axon growth occurs in two phases; the first, which is rapid and occurs between neuropiles, is followed by slow growth within the neuropiles. Differences are seen in the behaviour of the axon growth cone during these two phases.
5. Temporal sequences of axon growth are important in growth between neuropiles and give rise to spatial order of fibre bundles at the neuropile surface. This spatial order appears to be important in generating patterns of individual connections during the second phase of axon growth.

Unpublished work reported here was carried out during the tenure of a post-doctoral fellowship from the Australian National University. I should like to thank Professor G. A. Horridge for facilities in his laboratory. I am grateful to Drs C. M. Bate, S. Benzer, T. E. Hanson, J. Kien and P. M. J. Shelton for permission to quote unpublished work, to V. B. Meyer-Rochow for assistance with the German literature and to Dr J. Altman for critical reading of the manuscript.

References

Alverdes, F. (1924). Die Wirkung experimenteller Eingriffe, insbesondere der Blendung, auf den histologischen Bau des Insektengehirns. *Zeitschrift für Morphologie und Ökologie der Tiere*, **2**, 189–216.

Ando, H. (1957). A comparative study on the development of ommatidia in Odonata. *Science Reports of the Tokyo University of Education: Kyoiku Daigaku*, **8**, 174–216.

Bauer, V. (1904). Zur innern Metamorphose des Zentralnervensystems der Insecten. *Zoologische Jahrbücher Abteilung für Anatomie und Ontogenie der Tiere*, **20**, 123–52.

Becker, H. J. (1957). Über Röntgenmosaikflecken und Defektmutationen am Auge von *Drosophila* und die Entwicklungsphysiologie des Auges. *Zeitschrift für Induktive Abstammungs-und Vererbungslehre*, **88**, 333–73.

Bernard, F. (1937). Recherches sur la morphogénèse des yeux composés d'arthropodes. *Bulletin Biologique de la France et de la Belgique*, Supplement **23**, 1–162.

Bodenstein, D. (1950). The postembryonic development of *Drosophila*. In *Biology of Drosophila*, ed. M. Demerec, pp. 275–367. New York: Wiley.

(1953). Postembryonic development. In *Insect Physiology*, ed. K. D. Roeder, pp. 822–65. New York: Wiley.

(1959). Contributions to the problem of eye pigmentation in insects: Studied by means of intergeneric organ transplantations in Diptera. *Smithsonian Miscellaneous Collections*, **137**, 23–42.

(1963). Humoral conditions and cellular interactions in the development of the insect eye. In *Insect Physiology*, Proceedings of the Twenty-Third Annual Biology Colloquium, ed. V. J. Brookes, pp. 1–12. Corvallis: Oregon State University Press.

Bott, H. R. (1928). Bieträge zur Kenntnis von *Gryrinus natator substriatus* Steph. I. Lebensweise und Entwicklung. II. Der Sehapparat. *Zeitschrift für Morphologie und Ökologie der Tiere*, **10**, 207–306.

Braitenberg, V. (1967). Patterns of projection in the visual system of the fly. I. Retina–lamina projections. *Experimental Brain Research*, **3**, 271–98.

(1970). Ordnung und Orientierung der Elemente im Sehsystem der Fliege. *Kybernetik*, **7**, 235–42.

Burke, J. A. (1956). The differentiation of the compound eyes of *Doleschallia bisaltida* Cramer. *Anatomical Record*, **125**, 619–20.

Cajal, S. R. & Sánchez y Sanchez, D. (1915). Contribución al conocimiento de los centros nerviosos de los insectos. *Trabajos del Laboratorio de Investigaciones Biológicas de la Universidad de Madrid*, **13**, 1–164.

Campos-Ortega, J. A. & Strausfeld, N. J. (1972). The columnar organization of the second synaptic region of the visual system of *Musca domestica* L. *Zeitschrift für Zellforschung und Mikroskopische Anatomie*, **124**, 561–85.

Chevais, S. (1937). Sur la structure des yeux implantés de *Drosophila melanogaster*. *Archives d'Anatomie Microscopique*, **33**, 107–12.

Constantineanu, M. J. (1930). Der Aufbau der Sehorgane bei den im Süsswasser lebenden Dipterenlarven und bei Puppen und Imagines von *Culex*. *Zoologische Jahrbücher Abteilung für Anatomie und Ontogenie der Tiere*, **52**, 253–346.

Corneli, W. (1924). Von dem Aufbau des Sehorgans der Blattwespenlarven und der Entwicklung des Netzauges. *Zoologisches Jahrbücher Abteilung für Anatomie und Ontogenie der Tiere*, **46**, 573–608.

Dietrich, W. (1909). Die Facettenaugen der Dipteren. *Zeitschrift für Wissenschaftliche Zoologie*, **92**, 465–539.

Drescher, W. (1960). Regenerationsversuche am Gehirn von *Periplaneta americana*. *Zeitschrift für Morphologie und Ökologie der Tiere*, **48**, 576–649.

Eastham, L. E. S. (1930). The embryology of *Pieris rapae* – Organogeny. *Philosophical Transactions of the Royal Society of London*, B, **219**, 1–50.

Edwards, J. S. (1967). Some questions for the insect nervous system. In *Insects and Physiology*, ed. J. W. L. Beament & J. E. Treherne, pp. 163–74. Edinburgh & London: Oliver & Boyd.

(1969). Postembryonic development and regeneration of the insect nervous system. *Advances in Insect Physiology*, **6**, 97–137.

Elofsson, R. & Dahl, E. (1970). The optic neuropiles and chiasmata of Crustacea. *Zeitschrift für Zellforschung und Mikroskopische Anatomie*, **107**, 343–60.

Gaze, R. M. (1970). *The Formation of Nerve Connections*. London & New York: Academic Press.

Gaze, R. M., Chung, S. H. & Keating, M. J. (1972). Development of the retinotectal projection in *Xenopus*. *Nature, New Biology, London*, **236**, 133–5.

Gieryng, R. (1965). Veränderungen der histologischen Struktur des Gehirns von *Calliphora vomitoria* (L). (Diptera) während der postembryonalen Entwicklung. *Zeitschrift für Wissenschaftliche Zoologie*, **171**, 80–96.

Gottschewski, G. H. M. (1960). Morphogenetische Untersuchungen an in vitro wachsenden Augenanlagen von *Drosophila melanogaster*. *Wilhelm Roux Archiv für Entwicklungsmechanik der Organismen*, **152**, 204–29.

Günther, K. (1912). Die Sehorgane der Larve und Imago von *Dytiscus marginalis*. *Zeitschrift für Wissenschaftliche Zoologie*, **100**, 60–115.

Haas, G. (1956). Entwicklung des Komplexauges bei *Culex pipiens* und *Aëdes aegypti*. *Zeitschrift für Morphologie und Ökologie der Tiere*, **45**, 198–216.

Hanson, T. E. (1972). Neurogenesis in the eye and optic tracts of *Drosophila*. *Annual Report of the Division of Biology, California Institute of Technology*.

Hanson, T. E., Jiang, Y-H. & Lee, J-Y. (1972). Growth cone dynamics in the lamina of *Drosophila*. *Annual Report of the Division of Biology, California Institute of Technology*.

Hanson, T. E., Ready, D. F. & Benzer, S. (1972). Use of mosaics in the analysis of pattern formation in the retina of *Drosophila*. *Annual Report of the Division of Biology, California Institute of Technology*.

Heller, R. & Edwards, J. S. (1968). Regeneration of the compound eye in *Acheta domesticus*. *American Zoologist*, **8**, 786.

Hertweck, H. (1931). Anatomie und Variabilität des Nervensystems und der Sinnesorgane von *Drosophila melanogaster* (Meigen). *Zeitschrift für Wissenschaftliche Zoologie*, **139**, 559–663.

Heymons, R. (1895). *Die Embryonalentwicklung von Dermapteren und Orthopteren unter besondere Berucksichtigung der Keimblatterbildung*. Jena: Gustav Fischer.

Hinke, W. (1961). Das relative postembryonale Wachstum der Hirnteile von *Culex pipiens, Drosophila melanogaster* und *Drosophila*-Mutanten. *Zeitschrift für Morphologie und Ökologie der Tiere*, **50**, 81–118.

Holmgren, N. (1909). Termitenstudien. *Svenska Vetenskaps – Akademien. Handlingar*. n(4)s. **44**, 1–215.

Horridge, G. A. (1968). Affinity of neurones in regeneration, *Nature, London*, **219**, 737–40.

Horridge, G. A. & Meinertzhagen, I. A. (1970). The accuracy of the patterns of connexions of the first- and second-order neurons of the visual system of *Calliphora*. *Proceedings of the Royal Society of London*, B, **175**, 69–82.

Hyde, C. A. T. (1972). Regeneration, post-embryonic induction and cellular interaction in the eye of *Periplaneta americana*. *Journal of Embryology and Experimental Morphology*, **27**, 367–79.

Jacobson, M. (1970). *Developmental Neurobiology*. New York: Holt, Rinehart & Winston.

Jörschke, H. (1914). Die Facettenaugen der Orthopteren und Termiten. *Zeitschrift für Wissenschaftliche Zoologie*, **111**, 153–280.

Kopeć, S. (1922). Mutual relationship in the development of the brain and eyes of Lepidoptera. *Journal of Experimental Zoology*, **36**, 459–68.

Laschat, F. (1944). Die embryonale und postembryonale Entwicklung der Netzaugen und Ocellen von *Rhodnius prolixus*. *Zeitschrift für Morphologie und Ökologie der Tiere*, **40**, 314–47.

Lerum, J. E. (1968). The postembryonic development of the compound eye and optic ganglia in dragonflies. *Proceedings of the Iowa Academy of Science*, **75**, 416–32.

Lew, G. T.-W. (1933). Head characters of the Odonata. *Entomologica Americana*, **14**, 41–97.

Lüdtke, H. (1940). Die embryonale und postembryonale Entwicklung des Auges bei *Notonecta glauca* (Hemiptera–Heteroptera). *Zeitschrift für Morphologie und Ökologie der Tiere*, **37**, 1–37.

Meinertzhagen, I. A. (1971). *The First and Second Neural Projections of the Insect Eye*. Ph.D. Thesis, University of St Andrews.

(1972). Erroneous projection of retinula axons beneath a dislocation in the retinal equator of *Calliphora*. *Brain Research*, **41**, 39–49.

Morest, D. K. (1969). The differentiation of cerebral dendrites: a study of the post-migratory neuroblast in the medial nucleus of the trapezoid body. *Zeitschrift für Anatomie und Entwicklungsgeschichte*, **128**, 271–89.

Mouze, M. (1972). Croissance et metamorphose de l'appareil visuel des Aeschnidae (Odonata). *International Journal of Insect Morphology and Embryology*, **1**, 181–200.

Nordlander, R. H. & Edwards, J. S. (1968). Morphological cell death in the post-embryonic development of the insect optic lobes. *Nature, London*, **218**, 780–1.

(1969*a*). Postembryonic brain development in the monarch butterfly, *Danaus plexippus plexippus*, L. I. Cellular events during brain morphogenesis. *Wilhelm Roux Archiv für Entwicklungsmechanik der Organismen*, **162**, 197–217.

(1969*b*). Postembryonic brain development in the monarch butterfly, *Danaus plexippus plexippus* L. II. The optic lobes. *Wilhelm Roux Archiv für Entwicklungsmechanik der Organismen*, **163**, 197–220.

Ouweneel, W. J. (1970). Developmental capacities of young and mature, wild-type and *opht* eye imaginal discs in *Drosophila melanogaster*. *Wilhelm Roux Archiv für Entwickslungsmechanik der Organismen*, **166**, 76–88.

Panov, A. A. (1957). Bau des Insektengehirn während der postembryonalen Entwicklung. *Entomologicheskoe Obozrenie*, **36**, 269–84. (In Russian).

(1960). The structure of the insect brain during successive stages of post-embryonic development. III. Optic lobes. *Entomological Review*, **39**, 55–68. (Translation).

Parker, G. H. (1897). The retina and optic ganglia in decapods, especially in *Astacus*. *Mittheilungen aus der Zoologischen Station zu Neapel*, **12**, 1–73.

Perry, M. M. (1968). Further studies on the development of the eye of *Drosophila melanogaster*. I. The ommatidia. *Journal of Morphology*, **124**, 227–48.

Pflugfelder, O. (1936–7). Vergleichend - anatomische, experimentelle und embryologische Untersuchungen über das Nervensystem und die Sinnesorgane der Rhynchoten. *Zoologica, Stuttgart*, **34**(93), 1–102.

(1937). Die Entwicklung der optischen Ganglien von *Culex pipiens. Zoologischer Anzeiger*, **117**, 31–6.

(1947). Die Entwicklung embryonaler Teile von *Carausius* (*Dixippus*) *morosus* in der Kopfkapsel von Larven und Imagines. *Biologisches Zentralblatt*, **66**, 372–87.

(1958). *Entwicklungsphysiologie der Insekten*, Second Edition. Leipzig: Akademische Verlagsgesellschaft.

Plagge, E. (1936). Transplantationen von Augenimaginalscheiben zwischen der schwarz – und der rotäugigen Rasse von *Ephestia kühniella* Z. *Biologisches Zentralblatt*, **56**, 406–9.

Postlethwait, J. H. & Schneiderman, H. A. (1971). Pattern formation and determination in the antenna of the homoeotic mutant *Antennapedia* of *Drosophila melanogaster. Developmental Biology*, **25**, 606–40.

Power, M. E. (1943). The effect of reduction in numbers of ommatidia upon the brain of *Drosophila melanogaster. Journal of Experimental Zoology*, **94**, 33–72.

Richard, G. & Gaudin, G. (1960). La morphologie du developpement du système nerveux chez divers insectes – cas plus particulier des centres et des voies optiques. In *Symposium on the Ontogeny of Insects. Acta Symposii de Evolutione Insectorum Praha 1959*, ed. I. Hrdý, pp. 82–9. Prague: Czechoslovak Academy of Sciences.

Riek, E. F. (1970). Mecoptera. In *The Insects of Australia*, pp. 636–46. Carlton: Melbourne University Press.

Sánchez y Sanchez, D. (1916). Datos para el conocimiento histogénico de los centros ópticos de los insectos. Evolución de algunos elementos retinianos del *Pieris brassicae* L. *Trabajos del Laboratorio de Investigaciones Biológicas de la Universidad de Madrid*, **14**, 189–231.

(1918). Sobre el desarrollo de los elementos nerviosos en la retina del *Pieris brassicae* L. *Trabajos del Laboratorio de Investigaciones Biológicas de la Universidad de Madrid*, **16**, 213–78.

(1919*a*). Sobre el desarrollo de los elementos nerviosos en la retina del *Pieris brassicae* L. (Continuación). *Trabajos del Laboratorio de Investigaciones Biológicas de la Universidad de Madrid*, **17**, 1–63.

(1919*b*). Sobre el desarrollo de los elementos nerviosos en la retina del *Pieris brassicae* L. (Continuación). *Trabajos del Laboratorio de Investigaciones Biológicas de la Universidad de Madrid*, **17**, 117–80.

Satô, S. (1951). Development of the compound eye of *Culex pipiens* var. *pallens* Coquillett (Morphological studies on the compound eye in the mosquito, no. II). *Science Reports of the Tôhoku University*, 4th Ser. (Biology), **19**, 23–8.

(1953*a*). Structure and development of the compound eye of *Aedes* (*Finlaya*) *japonicus* Theobald (Morphological studies of the compound eye in the mosquito, no. III). *Science Reports of the Tôhoku University*, 4th Ser. (Biology), **20**, 33–44.

(1953*b*). Structure and development of the compound eye of *Anopheles hyrcanus sinensis* Wiedemann (Morphological studies on the compound eye in the mosquito, no. IV). *Science Reports of the Tôhoku University*, 4th Ser. (Biology), **20**, 45–53.

Schaller, F. (1960). Étude du développement post-embryonnaire d'*Aeschna cyanea* Mull. *Annales des Sciences Naturelles, Zoologie*, 12ᵉ série, **2**, 751–868.

Schoeller, J. (1964). Recherches descriptives et expérimentales sur la céphalogenèse de *Calliphora erythrocephala* (Meigen), au cours des développements embryonnaire et postembryonnaire. *Archives de Zoologie Expérimentale et Générale*, **103**, 1–216.

Schrader, K. (1938). Untersuchungen über die Normalentwicklung des Gehirns und Gehirntransplantationen bei der Mehlmotte *Ephestia kühniella* Zeller nebst einigen Bemerkungen über das Corpus allatum. *Biologisches Zentralblatt*, **58**, 52–90.

Seidel, F. (1935). Der Anlagenplan im Libellenei. *Wilhelm Roux Archiv für Entwicklungsmechanik der Organismen*, **132**, 671–751.

Shatoury, H. H. El. (1956). Differentiaton and metamorphosis of the imaginal optic glomeruli in *Drosophila*. *Journal of Embryology and Experimental Morphology*, **4**, 240–7.

(1963). The development of the 'eyeless' condition in *Drosophila*. *Caryologia*, **16**, 431–7.

Stein, I. (1954). Veränderungen am histologischen Bau der Sehzentren von Libellenlarven nach Blendung. *Österreichische Zoologische Zeitschrift*, **5**, 159–71.

Steinberg, A. G. (1941). A reconsideration of the mode of development of the bar eye of *Drosophila melanogaster*. *Genetics*, **26**, 325–46.

Strausfeld, N. J. (1970). Golgi studies on insects. Part II. The optic lobes of Diptera. *Philosophical Transactions of the Royal Society of London*, B, **258**, 135–223.

(1971*a*). The organization of the insect visual system (light microscopy). I. Projections and arrangements of neurons in the lamina ganglionaris of Diptera. *Zeitschrift für Zellforschung und Mikroskopische Anatomie*, **121**, 377–441.

(1971*b*). The organization of the insect visual system (light microscopy). II. The projection of fibres across the first optic chiasma. *Zeitschrift für Zellforschung und Mikroskopische Anatomie*, **121**, 442–54.

Strausfeld, N. J. & Blest, A. D. (1970). Golgi studies on insects. Part I. The optic lobes of Lepidoptera. *Philosophical Transactions of the Royal Society of London*, B, **258**, 81–134.

Tennyson, V. M. (1970). The fine structure of the axon and growth cone of the dorsal root neuroblast of the rabbit embryo. *Journal of Cell Biology*, **44**, 62–79.

Trujillo-Cenóz, O. & Melamed, J. (1966). Compound eye of dipterans: anatomical basis for integration – an electron microscope study. *Journal of Ultrastructure Research*, **16**, 395–8.

Umbach, W. (1934). Entwicklung und Bau des Komplexauges der Mehlmotte *Ephestia kühniella* Zeller nebst einigen Bemerkungen über die Entstehung der optischen Ganglien. *Zeitschrift für Morphologie und Ökologie der Tiere*, **28**, 561–94.

Viallanes, H. (1885). Études histologiques et organologiques sur les centres nerveux et les organes des sens des animaux articulés. Troisième mémoire.

Le ganglion optique de quelques larves de diptères (*Musca, Eristalis, Stratiomys*). *Annales des Sciences Naturelles* (*Zoologie*) (6), **19**, 1–34.
(1891). Sur quelques points de l'histoire du développement embryonnaire de la mante religieuse (*Mantis religiosa*). *Annales des Sciences Naturelles* (*Zoologie*) (7), **11**, 283–328.
Vogt, M. (1946). Zur labilen Determination der Imaginalscheiben von *Drosophila*. I. Verhalten verschiedenaltriger Imaginalanlagen bei operativer Defektsetzung. *Biologisches Zentralblatt*, **65**, 223–38.
Volkonsky, M. (1938). Sur la formation des stries oculaires chez les Acridiens. *Comptes Rendus des Séances de la Société de Biologie*, **129**, 154–7.
Waddington, C. H. (1962). Specificity of ultrastructure and its genetic control. *Journal of Cellular and Comparative Physiology*, Supplement **60**, 93–105.
Waddington, C. H. & Perry, M. M. (1960). The ultra-structure of the developing eye of *Drosophila*. *Proceedings of the Royal Society of London*, B, **153**, 155–78.
Weber, H. (1966). *Grundriss der Insektenkunde*, Fourth Edition. Jena: Gustav Fischer.
Wheeler, W. M. (1893). A contribution to insect embryology. *Journal of Morphology*, **8**, 1–160.
White, R. H. (1961). Analysis of the development of the compound eye in the mosquito *Aedes aegypti*. *Journal of Experimental Zoology*, **148**, 223–40.
White, R. H. (1963). Evidence for the existence of a differentiation center in the developing eye of the mosquito. *Journal of Experimental Zoology*, **152**, 139–48.
Wigglesworth, V. B. (1953). The origin of sensory neurones in an insect, *Rhodnius prolixus* (Hemiptera). *Quarterly Journal of Microscopical Science* **94**, 93–112.
Wolbarsht, M. L., Wagner, H. G. & Bodenstein, D. (1966). Origin of electrical responses in the eye of *Periplaneta americana*. In *The Functional Organization of the Compound Eye*, Wenner Gren Symposium 7, ed. C. G. Bernhard, pp. 207–17. Oxford: Pergamon.
Wolsky, A. (1938). Experimentelle Untersuchungen über die Differenzierung der zusammengesetzten Augen des Seidenspinners (*Bombyx mori* L.) *Wilhelm Roux Archiv für Entwicklungsmechanik der Organismen*, **138**, 335–344.
(1947). The growth and differentiation of retinula cells in the compound eyes of the silkworm (*Bombyx mori* L.) *Experimental Cell Research*, Supplement **1**, 549–54.
(1956). The analysis of eye development in insects. *Transactions of the New York Academy of Sciences*, Ser. II, **18**, 592–6.
Wolsky, A. & Wolsky, M. de I. (1971). Phase specific and regional differences in the development of the complex eye of the mulberry silkworm (*Bombyx mori* L.) after unilateral removal of the optic lobe of the brain in early pupal stages. *American Zoologist*, **11**, 679.
Yagi, N. & Koyama, N. (1963). *The Compound Eye of Lepidoptera*. Tokyo: Maruzen.
Young, E. C. (1969). Eye growth in Corixidae (Hemiptera: Heteroptera). *Proceedings of the Royal Entomological Society of London*, A, **44**, 71–8.

Zavřel, J. (1907). Die Augen einiger Dipterenlarven und -Puppen. *Zoologischer Anzeiger*, **31**, 247–55.

Zawarzin, A. (1913). Histologische studien über Insekten. IV. Die optischen Ganglien der *Aeschna-Larven. Zeitschrift für Wissenschaftliche Zoologie*, **108**, 175–257.

Note added in proof

Support for the notion that axonal outgrowth of lamina monopolar cells is induced by ingrowing retinula axons and direct observation of interneuropilar axon growth patterns comes from recent observations by Lopresti, Macagno & Levinthal (1973) on the developing optic lobe of *Daphnia*. In this small crustacean, axons from the five monopolar cells of a cartridge group are put out in the same sequence as their cell bodies are contacted by an incoming retinular axon bundle.

Lopresti, V., Macagno, E. R. & Levinthal, C. (1973). Structure and development of neuronal connections in isogenic organisms: Cellular interactions in the development of the optic lamina of *Daphnia*. *Proceedings of the National Academy of Sciences, USA*, **70**, 433–7.

Proliferation, movement, and regression of neurons during the postembryonic development of insects

R. L. Pipa

An adult insect is capable of behaviour which, in many respects, is strikingly different from that shown by the larva from which it has arisen. The divergence is made possible by an orderly development of new organs or organ parts from previously undifferentiated cells. The transformation is, of course, under genetic control. It is signalled and potentiated by changes in the hormonal milieu, and it is characterised by cellular proliferation, migration, growth, and differentiation. Where departure from the larval form is pronounced, extensive cell death also contributes to the process.

The precise way behavioural patterns are influenced by addition, deletion, or spatial redistribution of neurons during the post-embryonic period remains to be demonstrated. Nevertheless, there is considerable histological evidence that remodelling of the nervous system does occur, and it is to be expected that the changes are functionally important. The mechanisms involved have scarcely been studied in a systematic fashion, and experimental analyses are rare. Yet, enough is known to suggest that the phenomenon offers much of neurobiological interest.

Multiplication of neurons

Symmetric and asymmetric cell divisions

Mitotic events that result in the production of new neurons after eclosion resemble those seen in the embryo (Bauer, 1904; Schrader, 1938). Each primary formative cell (neuroblast) divides asymmetrically, with its spindle axis perpendicular to the neuropile surface. In this way a daughter neuroblast and a smaller ganglion mother cell are formed. The new neuroblast retains the position and function of its predecessor, whereas the ganglion mother cell, located closer to the neuropile, divides equally to produce preganglion cells.

In the monarch butterfly brain the spindle axis of each ganglion mother cell is usually oriented at about 45° to the neuropile surface (Nordlander & Edwards, 1969*a*). The ganglion mother cell divides an undetermined number of times. Its progeny become arranged in regularly aligned columns, with the more recently formed cells located adjacent to the neuroblast. It is from the preganglion cells that neurons and, perhaps, certain of the neuroglia differentiate.

This is not the only way neurons proliferate during the post-embryonic period. In certain cases daughter cells are not produced in regular columns headed by separate neuroblasts (Panov, 1960). For example, in pharate pupae of the cabbage butterfly, in honey bee larvae, and in dragonfly nymphs the optic lobe neuroblasts occur in two confluent groups. Neuroblasts of the more medial aggregation are larger than those of the lateral, and they divide asymmetrically as described above. The neuroblasts of the lateral aggregation, by contrast, only divide symmetrically. Furthermore, Panov (1960) finds that during the development of the optic lobes of the katydid and mayfly the ganglion mother cells and daughter neuroblasts arise solely by equal cell divisions. Symmetrically dividing neuroblasts also have been described in the developing corpora pedunculata ('mushroom bodies') of the monarch butterfly brain (Nordlander & Edwards, 1970). These observations imply that neuron proliferation may be occurring in regions where asymmetric cell divisions are not evident. It would be prudent to keep this in mind when reports claiming no evidence for neuron multiplication are being evaluated.

It is doubtful whether events leading to the production of neurons have been separated consistently and accurately from those involved in neuroglia replication. In early developmental stages the two kinds of cells cannot be identified readily, for their cytoplasm content is low and axons are vanishingly thin. Hopefully, distinctions will become clearer as more workers use silver staining techniques and electron microscopy in their studies.

Sources and mitotic activity of neuroblasts

Some of the large neuroblasts that divide asymmetrically are survivors from the embryonic population, while others appear to arise from undifferentiated cells during the larval stages. All of the large neuroblasts in the ventral nerve cord of the oak silkworm perish during blastokinesis, and in the brain only four pairs remain to form the association centres after the insect has hatched (Panov, 1963). The house cricket, *Acheta* (*Gryllus*) *domesticus*, also loses most of its large neuroblasts during the last stages of embryonic development; only those in the olfactory lobes, optic lobes, and corpora pedunculata are retained (Panov, 1966). In larval brains of the Mediterranean flour moth (Schrader, 1938), the oak silkmoth (Panov, 1963), and the monarch butterfly (Nordlander & Edwards, 1969*b*) the large neuroblasts held over from the embryonic stage appear to be supplemented by others formed *de novo* from smaller, undifferentiated cells.

In *Acheta* (Panov, 1962) and in the monarch butterfly (Nordlander & Edwards, 1969*a*) neuroblasts and ganglion mother cells divide continuously throughout the period from hatching to adulthood, and there is no apparent correlation between the frequency of proliferation and moulting cycles. This contrasts with the neuroglia, which like many other cells in the insect show mitotic activity that can be correlated with the stage of post-embryonic development. Interestingly, mitoses of neuroblasts and ganglion mother cells do not invariably cease once adulthood is reached; they have been found in adult brains of the katydid *Tettigonia viridissima* L. (Panov, 1960) and the milkweed bug *Oncopeltus fasciatus* (Johansson, 1957).

Estimates of numbers and kinds of neurons

Direct counts of the total number of cells in selected portions of the central nervous system are needed if the remodelling that occurs is to be defined. Though exceedingly useful, such studies are tedious and have seldom been reported.

An example of a precise counting technique is given by Becker (1965), who determined the numbers of neurons and glia in the second abdominal ganglion of the adult stick insect, *Carausius morosus*. He marked every identified cell on a photomicrograph of each histological section. To avoid counting the same cell more than once he tallied only those nuclei which contained nucleoli. Where nucleoli were not seen he counted those nuclei in which the maximum diameter was lying on the section. In this painstaking manner he determined a total neuron population of 608, and a ratio of one neuron to 1.5 glial cells.

Gymer & Edwards (1967) counted all neuronal and glial nuclei in each section through the terminal ganglion of *Acheta* and applied a correction formula to compensate for fragmentation due to sectioning. This ganglion originates by the embryonic fusion of the last four neuromeres (Panov, 1966). Gymer & Edwards concluded that the neuron population remains at about 2100 in all stages of postembryonic development, while the glial cells increase 17-fold. Their data indicated that the ratio of neurons to glia changes from 1:0.5 in the first instar to 1:8 in the young adult.

Sbrenna (1971) analysed the first abdominal ganglion of the locust *Schistocerca gregaria* in the same manner as Gymer & Edwards, and estimated that the glial cells increase from 500 in the newly-hatched hopper to 1800 by the 16th day of imaginal life. The neurons were said to remain at about 430 during the entire period, while the ratio of neurons to glial cells diminishes from 1:1 in the first instar to 1:4 in the adult.

In the brain large numbers of neurons can be added. Neder (1959) estimated that the complement of globuli cells in the corpora pedunculata of the cockroach *Periplaneta americana* rises from 73000 to 400000 during the entire postembryonic period. The increment is not constant from instar to instar, and it is greatest during earlier instars. The numbers of globuli cells in the corpora pedunculata reportedly increase 10- to 11-fold during the development of *Drosophila melanogaster*; there are about 250 in the first instar larva and about 2640 in the adult (Hinke, 1961).

It has been proposed that few new neurons are contributed to the ventral nerve cord of developing hemimetabolous insects. This view is supported by the data of Gymer & Edwards (1967) and Sbrenna (1971) cited above. It is also upheld by Panov (1966), who found that neuroblasts disappear from abdominal and thoracic ganglia of *Acheta* during its embryonic development. In the mole cricket, *Gryllotalpa gryllotalpa*, however, neuroblasts are present in the ventral nerve cord until the end of the first larval instar (Panov, 1966). Johansson (1957) likewise concluded that neuroblasts and dividing neurons are not to be found in the developing suboesophageal and first thoracic ganglion of *Oncopeltus*. It is uncertain to what extent these last two investigators have relied on the questionable assumption that neuron proliferation must be accompanied by asymmetrical cell divisions.

In thoracic ganglia of holometabolous insects, too, the incidence of neuron proliferation needs to be assessed. Heywood (1965) observed asymmetrical divisions of neuroblasts in thoracic ganglia of cabbage butterfly larvae, but none in pupae. In the latter stage he noted much mitotic activity in 'small neural cells'. Though he supposed that these were ganglion mother cells which had been carried over from the larval instar, his reasons for doing so are not clear.

Within the first 120 h after larval–pupal ecdysis of the wax moth, *Galleria mellonella*, asymmetrical cell divisions have not been found in prothoracic and mesothoracic ganglia (Pipa, unpublished). Yet, during that time the complement of interneurons in connectives between these centres increases from about 1000 to 7000 (Tung & Pipa, 1972). Autoradiographic studies that are in progress reveal most of the mitotic activity in small cells bordering the neuropile. The difficult question that needs to be resolved is whether these are glial cells, as their location suggests, or whether some may be neurons that are developing. The likelihood that the interneurons originate in ganglia outside the thorax must first be tested. Barring that possibility, it is conceivable that they arise from thoracic preganglion cells produced during earlier stages, only sending forth their processes at metamorphosis.

While it is certain that many interneurons are added to the insect nervous system after eclosion, it is unknown whether motor neurons are produced. In relationship to this, a consideration of developing effectors may be instructive.

Those muscles that differentiate in the embryo continue to grow during larval life. In the cockroach *Periplaneta americana*, where it has been examined closely (Teutsch-Felber, 1970), this growth is accomplished by an increase in the numbers and size of muscle fibres. Since an entire muscle is served by only one or two neurons, proliferation of muscle fibres must be met by further branching of existing axons. New motor neurons would not be expected in those cases.

The situation may be quite different where myoblasts, not differentiated muscles, are present in the larvae (Snodgrass, 1954; Daly, 1964). These develop at metamorphosis, and it is uncertain whether motor connections are established during embryogenesis, or whether they are formed after eclosion.

In other instances, muscles that are formed postembryonically replace groups that functioned at an earlier stage. The new muscles seem to arise from the nuclei that remain after their predecessors have regressed (Nüesch, 1968; Scudder & Hewson, 1971; Wigglesworth, 1956). For differentiation to occur, an intact nerve supply is sometimes necessary. There are indications that the old muscle and the new are served by the same axon, but this should be restudied.

By examining histological sections of ganglia stained by routine methods it is difficult, if not impossible, to determine whether the numbers of motor neurons increase. Special procedures are required to demonstrate that a centrally located nerve cell sends a process out along a nerve to an effector organ. The mapping of motor components during development may be facilitated by the recently available techniques for motor neuron identification, which are discussed by Bentley and by Young in this volume. In any case it is likely to be a difficult undertaking. The view, adopted by some investigators, that motor neurons are readily separable from interneurons by the larger size of their perikarya requires verification. At this time it should be dismissed as being unduly optimistic.

Careful analyses with the electron microscope would be expected to provide direct information about changing patterns of afferent and efferent innervation. To date, few studies of this kind have been made. We know that the numbers of axons in peripheral nerves may increase during metamorphosis (Osborne, 1966; Randall & Pipa, 1969), but the proportion of motor to sensory units in nerves at different times remains to be determined. In *Acheta* the total axon population of a cercal nerve rises from 300 in the first instar to 10145 in the adult (Edwards, 1969*a*) Though the origins of most of these axons are thought to be peripheral, their connections have yet to be defined.

Differentiation of the neuropile

Because synapses are largely, if not entirely, restricted to the neuropiles, the development of these regions is of particular interest. The sensory association centres of the brain have received the greatest attention, for their structure has seemed more ordered than that of other neuropiles. This regularity has facilitated gross comparisons between homologous brain regions, but it has not led to an immediate recognition of neural pathways. In most cases we are unable to identify major neuronal connections, let alone tell precisely how these change during development. Part of the failure is due to the inappropriate histological methods that have been employed; most workers have been content to study sections stained with iron haematoxylin. The intensive application of degeneration techniques, silver impregnation, and electron microscopy during the time these centres are developing should do much to unravel their complex organisation.

Edwards (1969*b*) presented a good review of the morphogenesis of each sensory association centre, so that will not be dealt with here. Instead, I shall summarise some interrelationships of a general kind.

Early differentiation of brain centres in the Hemimetabola

Those few studies that have been published of brain development in hemimetabolous insects show that the adult form of most major sensory association centres is already present in newly-hatched larvae. The generalisation which seems to be emerging is that amongst these insects the distinguishing outlines of adult brain centres are formed in the embryo, not in the larva. Perhaps that is to be expected; the larval sensory organs will also serve the adult and, except for an increase in the size and number of sensilla, they undergo little structural change.

The presence of definable association centres in these newly-hatched larvae should not be taken to mean that scant differentiation occurs later. As we have seen, neuron proliferation does occur. Moreover, during larval instars the neuropiles of some of these centres increase in volume dramatically (Afify, 1960; Johansson, 1957; Neder, 1959). To an undetermined extent that growth must reflect the formation of new connections.

A few exceptions to patterns of early brain development amongst the Hemimetabola have been noted, and probably more will be discovered. With regard to brain differentiation, a newborn mole cricket corresponds to a house cricket embryo (Panov, 1966). The calyx of its corpora pedunculata and the adult shape of its central body are initially seen in the second larval instar. The termite, *Calotermes flavicolis*, also shows retarded brain development. Not before the fifth larval instar, when the ommatidia become differentiated, are three optic glomeruli evident (Richard, 1950).

Correlated development of eyes and optic centres in the Holometabola

By comparison with the Hemimetabola, differentiation of brain centres amongst immature holometabolous insects appears to be less uniform, and attempts have been made to relate this to the extent of sensory receptor development that is completed during embryogenesis. Panov (1960), for example, suggested that those larvae with well-developed visual organs at the time of hatching (e.g. sawflies (Hymenoptera), campodeiform larvae of Neuroptera and Coleoptera) possess three optic glomeruli. When stemmata (larval photoreceptor cells) are less well developed (lepidopteran larvae, eruciform larvae of Coleoptera) the number is reduced to one or two. In blind larvae (e.g. honey bees) no optic centres are to be found in the first instar. This correlation can also be observed amongst the Diptera (Hinke, 1961). First instar larvae of the mosquito, *Culex pipiens*, have well-developed eyes. They are also endowed with three optic glomeruli. Larvae of wild-type *Drosophila melanogaster*, being blind, lack a medulla externa and lobula (medulla interna) until the third larval instar.

While the extent of embryonic differentiation of the optic glomeruli may be correlated with the presence or absence of developing eyes, it is clear that during the postembryonic period these centres begin to differentiate before the ommatidia do. Three optic glomeruli are evident in the 5-day-old honey bee larva (Panov, 1960; Lucht-Bertram, 1962). At that time the medulla externa and lobula are connected by the inner chiasma. In the monarch butterfly the optic centre neuropiles arise during the larval stadia, but the growth of one of these, the lamina ganglionaris, is most rapid after pupation when the ommatidia differentiate (Nordlander & Edwards, 1969*b*). The medulla externa and lobula of wild-type *Drosophila* (Hinke, 1961) and the blowfly *Calliphora* (Gieryng, 1965) also form in the larva. They increase in volume most strikingly at metamorphosis, and at about that time the lamina ganglionaris first becomes visible.

Influence of peripheral fields on neuropile differentiation

The correlations summarised above relate to an important question: how does the postembryonic development of peripheral fields affect the emergence of central connections? This has been investigated by examining neuropiles that have differentiated in the absence of selected peripheral organs, a condition arising naturally in neurogenetic mutations (Power, 1943, 1946; Hinke, 1961) or achieved by ablation (Alverdes, 1924; Chiarodo, 1963; Panov, 1961). A correspondence between the size of the peripheral field and the volume of the neuropile that develops has been revealed by both methods, but the interactions are complex and their precise nature is unknown.

Power (1943) demonstrated that genetic reduction in numbers of ommatidia causes hypoplasia of the optic glomeruli of *Drosophila* and changes in their histology. The volume of each glomerulus varies linearly with ommatidial number, but in eyeless mutants only the lamina ganglionaris is absent; the other glomeruli demonstrate a reduction of 57–85 %. His data indicate that the hypoplasia reflects the ingrowth of a smaller number of axons from the genetically reduced disc, and that it is not a primary genetic response by the brain. When flies have a large eye on one side and a reduced eye on the opposite side the volume of a glomerulus is correlated only with the facet number of the adjacent side. Also, in those animals in which ommatidia are present but unconnected to the brain the volumes and histology of the optic glomeruli are like those of eyeless individuals. Hinke (1961), who analysed the allometric growth of the brain centres of *Drosophila* eye mutants, confirmed this relationship between the extent of eye reduction and the volume of the optic glomeruli.

In a subsequent investigation, Power (1946) made the curious observation that the volume of the antennal glomeruli also diminishes linearly with reduction in the size of the compound eyes. The hypoplasia is asymmetric in flies with unilateral eye reduction, suggesting a transsynaptic developmental influence (Jacobson, 1970, review) which works specifically between the two sensory centres of one side. Although eyelessness is associated with a 27 % hypoplasia in the antennal glomeruli, Power was unable to detect a corresponding disturbance in the morphology of these centres.

Hinke (1961) also noted a reduction in the volume of antennal glomeruli of *Drosophila* eye mutants: the olfactory glomeruli of wild-type flies occupy 3.0 % of the total brain volume minus the optic lobes, while the value is 1.7 % for eyeless. In addition, he reported that the relative size of the corpora pedunculata parallels the state of eye development.

Chiarodo, Reing & Saranchek (1971) estimated the numbers of cells in the thoracico-abdominal nerve centres of several winged but flightless *Drosophila* mutants. They were able to confirm the original conclusion of Power (1950) that inability to fly is not correlated with hypoplasia within that centre. Interestingly, they discovered a numerical hyperplasia of ganglion cells in bithorax and vestigial mutants. Because the bithorax mutant is characterised by a tendency of the metathorax and first abdominal segments to resemble the mesothorax they suggest that the hyperplasia might be related to an increased peripheral field of innervation. This explanation would not apply to the hyperplasia in the vestigial mutant, however, and in neither case are the mechanisms involved known.

The likelihood that neuropile morphogenesis can be influenced by surgically removing effector organs has received little attention. In the single investigation of this possibility (Chiarodo, 1963) the consequences of deafferentation (Alverdes, 1924; Panov, 1961) were not excluded. Chiarodo removed the mesothoracic leg discs from blowfly larvae and found that the volume of the corresponding neuropile in the adults was reduced 37%, but no qualitative changes were observed. There was a 27% decrease in the number of small cells in the mesothoracic cortex, while the giant cell populations remained unaffected. Chiarodo concluded that the small cells represent the leg motor neurons, and that some of the neuropile hypoplasia reflects their retrograde degeneration. He did not mention the possibility that some, if not all, of these small cells might be interneurons that degenerate because of a lack of transsynaptic stimulation by sensory receptors of the leg.

In insects, as in many vertebrates (Jacobson, 1970), intracentral neurogenesis can occur in the absence of peripheral connections. Thus Edwards & Palka (this volume) have shown that the post-embryonic development of the giant fibres of the ventral nerve cord in crickets continues normally in the absence of sensory input from the cercal sensilla.

Nevertheless, differentiating receptor organs assert a substantial influence: they induce an increase in neuropile volume during postembryonic development. This effect has been characterised quantitatively to a first approximation only, and it remains to be defined in qualitative terms. To what extent does the volumetric effect involve neurons, and to what extent glia? Are interneurons affected? Does neuropile hypoplasia reflect a failure of cells to proliferate, or is it caused by premature cell death? These are but a few of the difficult basic questions that must be answered if the mechanisms involved are to be understood.

Migration of neurons

As nerve fibres grow into the neuropile and establish new synaptic contacts the spatial distribution of neuron perikarya in the cortex changes. Does this mean that ganglion mother cells arise at one locus and migrate to another before sending forth their processes? Nordlander & Edwards (1969*b*) do not favour that interpretation. Instead, their autoradiographic studies suggested that developing neurons become passively displaced by adjacent differentiating fibres, by cell enlargement, and by accelerated neuropile growth. They admit, however, that their data do not eliminate the possibility that neuron migration may play a role, so the question is unresolved.

There are two unambiguous examples of neuron migrations occurring during insect postembryonic development: (1) the retraction of larval photoreceptor cells from the corneal lens, followed by their movement towards the brain, and (2) the shortening of the ventral nerve cord that may result in the fusion of ganglia. The first phenomenon happens during the metamorphosis of certain Diptera (Dietrich, 1909), Coleoptera (Günther, 1911; Ho, 1961; Murray & Tiegs, 1935), Hymenoptera (Corneli, 1924) and Lepidoptera (Johansen, 1893; Nordlander & Edwards, 1969*a*; Schrader, 1938). As a consequence of centripetal migration the stemmata, attached to their peripheral nerves and shrouded by pigment, may be found at variable distances from the compound eyes of the full-grown imago. The assumption that these 'rudimentary' organs no longer function as photoreceptors should be avoided, since neither their fine structure nor physiology has been investigated. The second example of neuron migration has been explored more intensively.

In the course of insect development the numbers of separate ganglia are usually, if not always, reduced. During embryogenesis the neuromeres of composite centres originate in their respective segments and then coalesce. In the most extreme cases the fusion may result in the formation of one (e.g. *Phormia*, Schaefer, 1938; *Drosophila*, Poulson, 1950) or two (e.g. *Oncopeltus*, Springer, 1967) ganglia composed of all the central neurons of the thorax and abdomen. During the postembryonic period, usually at metamorphosis, striking differences in the number of separate ganglia are also caused by dissimilarities in the extent of nerve cord shortening (Brandt, 1879*a–f*; Cody & Gray, 1938; Menees, 1961;

Tiegs, 1922; Woolley, 1943). Because the ultimate consequences of nerve cord shortening during the two stages are identical it is tempting to speculate that the developmental mechanisms may be similar. This remains to be demonstrated, however, for only the shortening that occurs at metamorphosis has been studied systematically.

The extent of ganglion fusion that accompanies the metamorphosis of Lepidoptera has been described by several investigators (Brandt, 1879*e*; Heywood, 1965; Newport, 1832, 1834; Pipa, 1963). In *Galleria* the transformation was measured at closely timed intervals (Pipa, 1963). At 33 °C the interganglionic connectives begin to shorten about one day before the pupa casts the larval skin, and they have shortened completely by about two days after that event. The meso- and metathoracic ganglia become contiguous and the metathoracic and first two abdominal ganglia fuse. Abdominal ganglia six to eight also merge to form a terminal abdominal centre, but the other abdominal connectives shorten less than 20 % of their initial lengths. The fusion of ganglia has little effect on the final length of the nerve cord, for much of the shortening is nullified by subsequent elongation and the adult cord is only 15–20 % shorter than that of the larva.

Mechanisms of nerve cord shortening

The histology of the connectives provides some clues about the mechanisms that might be involved. The larval connectives consist of 800–1000 axons, most of which are isolated from each other by neuroglial cytoplasm (Pipa & Woolever, 1965; Tung & Pipa, 1972). The glial sheaths of each connective are contributed by 3–5 giant cells with flattened polyploid nuclei located at the periphery, and by numerous Schwann cells (Pipa & Woolever, 1964). As the connectives shorten their histology changes markedly. Schwann cell nuclei previously restricted to the connective extremities now become numerous along its length. Individual axons are no longer surrounded by glial sheaths; instead, bundles of 'naked', contiguous axons are seen for the first time, and these are separated from one another by wide expanses of glial cytoplasm and enlarged extracellular spaces. The axons within shortening connectives progressively become looped. After shortening is completed the loops gradually diminish until, within a few days before adult emergence, they have disappeared.

The occurrence of looped axons in shortening connectives suggests a build-up of intrinsic tractive forces. It seems unlikely that those forces result from uniform contraction of axoplasm proteins, for looping would hardly be expected if such were the case. Instead, we favour the view that the ganglia are drawn together by the contraction or migration of glial cells. The increase in numbers of Schwann cells along the connectives, the disappearance of glial wrappings from about individual axons, and the occurrence of wide expanses of glial cytoplasm and enlarged extracellular spaces provide some support for this interpretation.

Heywood (1965) examined the histology of shortening interganglionic connectives during the metamorphosis of cabbage butterfly larvae, but failed to detect axon loops. He suggested that axon shortening is accomplished by 'reabsorption' of axoplasm, and that this results in connective shortening. It should be noted, however, that he studied histological sections stained with haematoxylin. In *Galleria*, at least, this stain is unsatisfactory for demonstrating looped axons. That axon looping is associated with connective shortening in another insect order, the Hymenoptera, was shown by Schwager-Hübner (1970), who described the fusion of thoracic and abdominal ganglia during the metamorphosis of *Apis*.

The histological data cited above are suggestive, but before definitive statements about mechanisms can be made experiments must be performed. In this regard, the capacity of interganglionic connectives to shorten after being transplanted was investigated (Pipa, 1967). When nerve cord segments from last instar larvae or early pupae are implanted into larvae they will shorten when the recipient pupates. There is convincing evidence that the shortening which takes place *ex situ* is not artifactitious: each transplant shortens in near-synchrony with the recipient's counterpart, and the histological changes which occur within the two are very similar. The likelihood that tractive forces are exerted on the segments by way of peripheral ganglionic connections is excluded by these results. Instead, the probability is enhanced that the ganglia are drawn together by mechanical forces originating within the connectives. Furthermore, when the ganglia are destroyed by cautery and the axons are given sufficient time to degenerate before the host pupates, shortening is not prevented. The connectives, like those of noncauterised implants, shorten together with the recipient's counterpart. The tractive forces must be exerted by the cells that remain, and the most probable candidates are the neuroglia.

Hormonal stimulation of nerve cord shortening

Larval connectives implanted into pupae less than two days old shorten completely; those implanted into larvae or into recipients more than two days beyond pupal ecdysis do not. When connectives that have begun to shorten are implanted into larvae they cease to shorten; when they are implanted into pupae less than two days old they shorten entirely. The endocrinological basis for these responses was explored by performing extirpation–implantation experiments, and by administering hormones to connectives *in vivo* (Pipa, 1969) and *in vitro* (Robertson & Pipa, 1973).

When the brain plus all acknowledged endocrines except the prothoracic glands are extirpated from an early pupa, connective shortening will cease. The effect can be remedied by implanting half a brain. If the prothoracic glands are deleted by removing the entire body anterior to the mesothorax shortening will also be prevented. As many as three brains implanted into a posterior fragment prepared this way will not reinitiate the process. These data indicate that the brain does not regulate connective shortening directly. Instead, it appears to function in the usual manner: a diffusible brain factor, acting on the prothoracic glands, initiates the production of the moulting hormone ecdysone. Direct proof that this steroid hormone is involved can be obtained by injecting it into posterior pupal fragments. Under these conditions α-ecdysone will stimulate connective shortening, and a dosage–response relationship can be demonstrated. An individual dosage of 0.08 μg will cause 50% of the meso–metathoracic connectives of such pupae to shorten at least half their initial length within three days.

The ability of ecdysones to stimulate connective shortening in the absence of all other tissue has been revealed by our recent studies of the system *in vitro* (Robertson & Pipa, 1973). The fraction of the connectives that shorten to at least half their mean initial length is directly related to the amount of β-ecdysone in the tissue culture medium. β-Ecdysone is about 140× more active than α-ecdysone, and the threshold of response is between 0.05 and 0.10 μg of β-ecdysone per millilitre of media. We are currently analysing the system using time-lapse cinemicrography.

Regression of neurons

Many larval muscles regress during the metamorphosis of holometabolous insects. Often, degeneration is incomplete and the muscles may become reconstituted to serve the imago (Snodgrass, 1954). In those instances it is conceivable, though unproved, that the old muscles and the new are supplied by the same motor neurons. Quite different consequences would be anticipated where replacement does not occur (cf. Daly, 1964; Randall, 1968), for then the motor neurons would probably degenerate as well. Likewise, those sensory neurons and interneurons that form neural connections peculiar to the larva would be expected to disappear during adult development.

The data that support these intuitive expectations are tenuous; in no instance has the progressive degeneration of mapped neurons been traced to completion or quantified. Because the criteria used to identify regressing neurons have not been defined rigorously, reports on the extent of degeneration are difficult to evaluate. Can we be certain that regression is entire, or that moribund cells are being described, when the complete sequence of changes in specific neurons has not been followed? It is well to keep these shortcomings in mind when accounts of neuron degeneration are being considered.

Signs of regression have been found in neurons that were thought to be fully differentiated (Heywood, 1965; Panov, 1961, 1963), as well as in neuroblasts and ganglion mother cells (Bauer, 1904; Nordlander & Edwards, 1968; Schrader, 1938). Scattered amongst the 'normal' neurons within the transforming olfactory lobes of the flour beetle, *Tenebrio molitor*, are structureless, chromophilic globules, and neurons with condensed, deeply-staining nuclei and cytoplasm (Panov, 1961). This was regarded as evidence for a progressive degeneration of larval neurons. During the metamorphosis of the oak silkworm substantial numbers of differentiated thoracic neurons also appear to degenerate, but the majority persist (Panov, 1963). Condensed, deeply-staining nuclei in ganglion mother cells of the optic lobes of the monarch butterfly are thought to be reliable signs of impending cell death (Nordlander & Edwards, 1968). Preliminary counts showed that the maximum number of 'degenerations' followed the maximum number of ganglion mother cell divisions by three or four days. It was conjectured that an excess of neurons is produced, and that those which fail to establish 'proper connections' in the neuropile perish.

Sensory association centres are remodelled during metamorphosis, and this may involve the degeneration of central connections. The larval optic centre neuropiles of the sawfly, *Pristophora pallipes*, are said to degenerate in the prepupa simultaneously with regression of the stemmata (Panov, 1960). Similarly, after the larval antennae of *Tenebrio* have degenerated the olfactory glomeruli change in appearance; they become less fibrous, and contain chromophilic inclusions which increase in number and size (Panov, 1961). The process continues throughout half the adult developmental period, and there is an overall contraction of the neuropile as the large granules disappear. The ultrastructural features of these changes have not been reported, but they would be of considerable interest.

Certain interganglionic connectives of *Galleria* shorten an average of 1.4 mm during metamorphosis (Pipa, 1963), and the axons become tightly looped. The loops persist several days after the ganglia meet, but they disappear before the adult emerges (Pipa & Woolever, 1964). The possibility that the loops are lost by autolysis and that some axons regress entirely is suggested by their fine structure (Tung & Pipa, 1972). As axons undergo metamorphosis they bear a striking resemblance to those that have been induced to degenerate by severing them from their cell bodies (Tung & Pipa, 1971). The axons are filled with spindle-shaped vesicles, apparently derived from neurotubules (Plate 1), and dense bodies, myeloid bodies, and autophagic vacuoles become abundant. There is an increase in the number of 'empty' axons with discontinuous cytomembranes, and these seem to be in the final stage of degeneration. For those instances when axons with interrupted limiting membranes also contain intact neurotubules (Plate 2) that interpretation may not be correct, and the significance of the change is unknown.

The fine structure of the polyploid neuroglia within transforming connectives suggests that these cells regress, too. Glial

---

Plate 1. Transverse section through a connective from an early pharate pupa of *Galleria*, showing an axon filled with small vesicles, a myeloid body M and an autophagic vacuole AV (From Tung & Pipa, 1972).

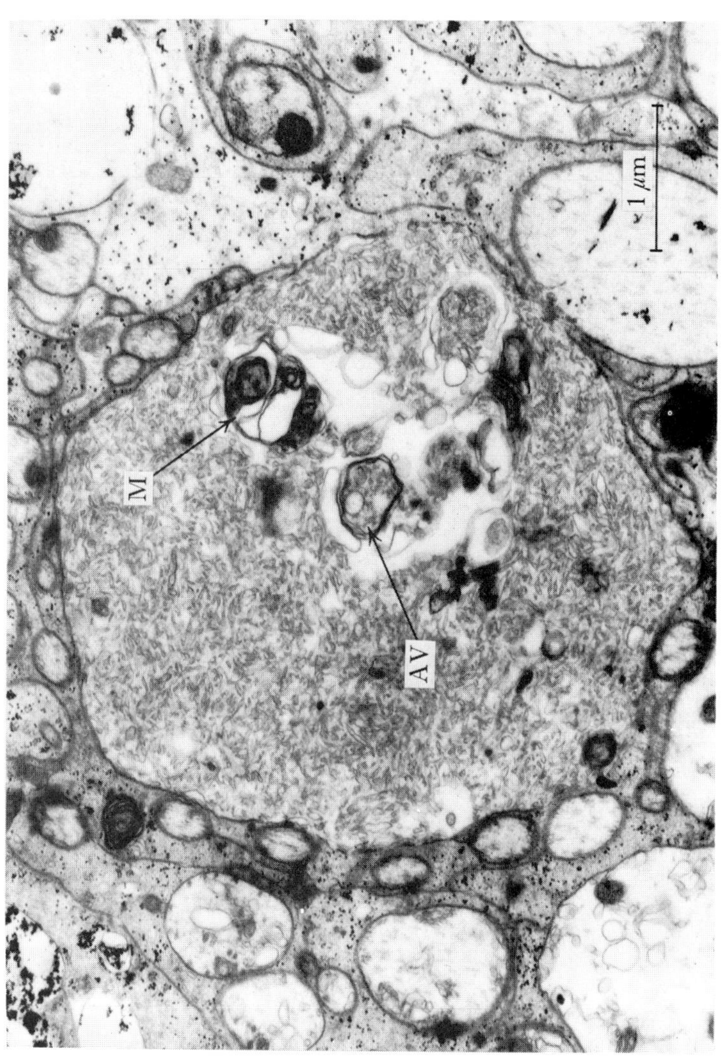
1 μm
M
AV

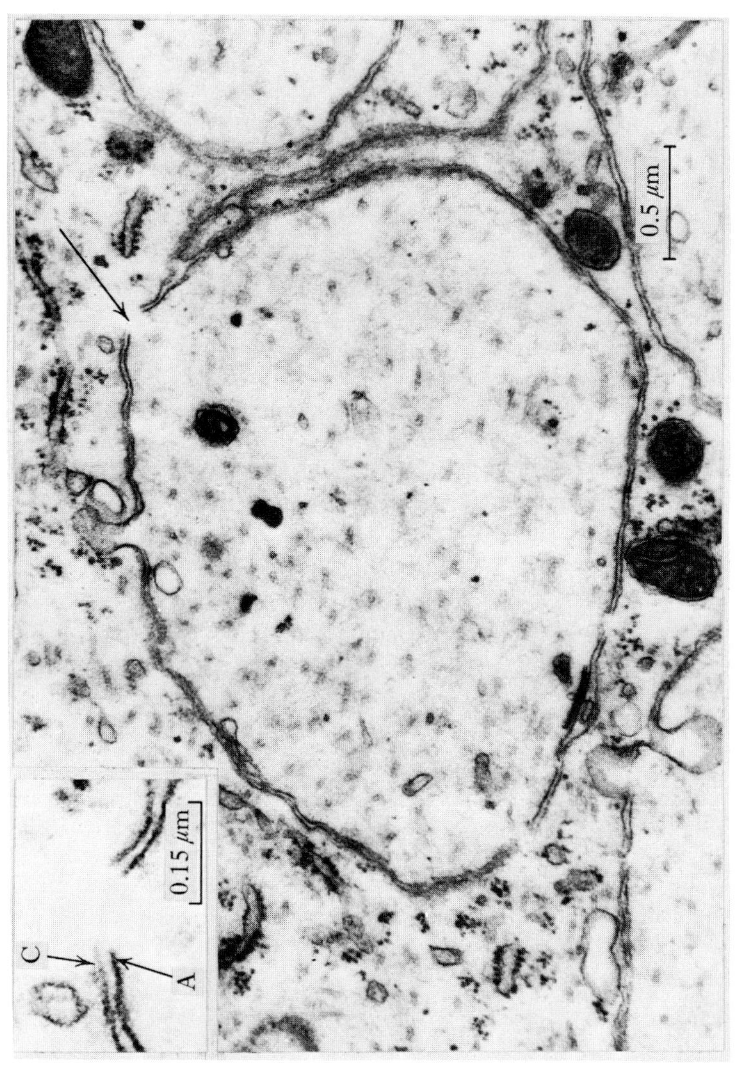
0.5 μm
0.15 μm
C
A

microtubules disappear, and vast, rarefied areas occur next to zones with the usual organelles. As in the axons, myeloid bodies, altered mitochondria, autophagic vacuoles, and interrupted cytomembranes are seen for the first time. Where the glial cytomembrane and axolemma both disappear the glial cytoplasm, extracellular channel and axoplasm become confluent (Plate 2). It may be that the glial cytoplasm, so abundant in larval connectives where it isolates nearly every axon, is reduced by autolysis. This could contribute to the adult condition, where multitudes of unsheathed, contiguous axons occur in irregular bundles bounded by scanty glial cytoplasm.

Conclusions

There is abundant evidence that the insect nervous system continues to develop throughout much of the postembryonic period. During that time neurons demonstrate characteristic patterns of proliferation, movement and degeneration, and increases in neuropile volume have also been documented. Neural differentiation in holometabolous insects is accelerated at the onset of metamorphosis, but it can hardly be considered restricted to that interval. Though less dramatic, the neurogenic events that occur during the larval instars of holometabolous and hemimetabolous insects are substantial. This more gradual pace of development would be expected to provide excellent opportunities to study how new units are integrated.

Only the broad outlines of neuropile development have been traced, and descriptions which are more incisively qualitative and quantitative are needed. Nowhere is the necessity for critical histological analyses more apparent than in this field, nor are such data apt to be as difficult to obtain. Until we have sound estimates of the proportions of sensory axons, interneurons, motor neurons and glia that are present the significance of volumetric changes during normal growth or after experimental manipulations will remain enigmatic. There is little hope of defining those environ-

---

Plate 2. Axolemmata and glial cytomembranes are interrupted in this connective at 120 h after pupal ecdysis. The extracellular space is lost, and glial cytoplasm and axoplasm are confluent (arrow), yet the presence of neurotubules within the axon indicates that it may not be degenerating. Clear separation of axolemma A and glial cytomembrane C by extracellular space is shown in the inset.

mental and genetic factors that influence the ontogeny of neural connections unless the units involved have been mapped.

The criteria for determining whether neuron proliferation is occurring should be reappraised. This seems particularly worthwhile in the light of reports that asymmetrical cell divisions do not invariably occur where ganglion mother cells are being produced. It could be that dividing neuroblasts or ganglion mother cells have been misinterpreted as neuroglia. Perhaps by injecting tritiated thymidine at a time when neuroblast divisions are suspected and searching for differentiated neurons with labelled nuclei, particularly in thoracic and abdominal ganglia, unexpected sources of neurons might be revealed.

Too few investigations of the influence of developing peripheral fields on intracentral neurogenesis have been conducted to allow meaningful generalisations. For the optic centres at least, it is clear that elements of neuropile morphogenesis can occur though the receptor organ is absent. This is shown during normal postembryonic development, when optic glomeruli begin to form in blind larvae. It is also evident in the case of adult *Drosophila* eyeless mutants, where the medulla externa and lobula are present, though less voluminous. The conclusion to be drawn from these observations is hardly surprising: information required for the establishment of connections is programmed within the central neurons or glia, and signals from sensory receptors are not essential. This is not to suggest that peripheral influences are totally inconsequential, however, for the hypoplasia of optic glomeruli of eyeless mutants has not been characterised adequately.

The research from our laboratory reported in this article was supported in part by USPHS Grant NS 03845 from the National Institute of Neurological Diseases and Stroke.

References

Afify, A. M. (1960). Über die postembryonale Entwicklung des Zentralnervensystems (ZNS) bei der Wanderheuschrecke *Locusta migratoria migratorioides* (R. u. F.) (Orthoptera–Acrididae). *Zoologische Jahrbücher Abteilung für Anatomie und Ontogenie der Tiere*, **78**, 1–38.

Alverdes, F. (1924). Die Wirkung experimenteller Eingriffe, insbesondere der Blendung, auf den histologischen Bau des Insektengehirns. *Zeitschrift für Morphologie und Ökologie der Tiere*, **2**, 189–216.

Bauer, V. (1904). Zur innern Metamorphose des Centralnervensystems der

Insecten. *Zoologische Jahrbücher Abteilung für Anatomie und Ontogenie der Tiere*, **20**, 123–52.

Becker, H. W. (1965). The number of neurons, glial and perineurium cells in an insect ganglion. *Experientia*, **21**, 719.

Brandt, E. (1879*a*). Vergleichend-anatomische Skizze des Nervensystems der Insekten. *Horae Societatis Entomologicae Rossicae*, **15**, 1–19.

(1879*b*). Über die Metamorphosen des Nervensystems der Insecten. *Horae Societatis Entomologicae Rossicae*, **15**, 20–30.

(1879*c*). Vergleichend-anatomische Untersuchungen über das Nervensystem der Hymenopteren. *Horae Societatis Entomologicae Rossicae*, **15**, 31–50.

(1879*d*). Vergleichend-anatomische Untersuchungen des Nervensystems der Käfer (Coleoptera). *Horae Societatis Entomologicae Rossicae*, **15**, 51–67.

(1879*e*). Vergleichend-anatomische Untersuchungen das Nervensystem der Lepidopteren. *Horae Societatis Entomologicae Rossicae*, **15**, 68–83.

(1879*f*). Vergleichend-anatomische Untersuchungen über das Nervensystem der Zweiflügler (Diptera). *Horae Societatis Entomologicae Rossicae*, **15**, 84–101.

Chiarodo, A. J. (1963). The effects of mesothoracic leg disc extirpation on the postembryonic development of the nervous system of the blowfly, *Sarcophaga bullata*. *Journal of Experimental Zoology*, **153**, 263–77.

Chiarodo, A., Reing, C. M. Jr. & Saranchak, H. (1971). On neurogenetic relations in *Drosophila melanogaster*. *Journal of Experimental Zoology*, **178**, 325–30.

Cody, F. P. & Gray, I. E. (1938). The changes in the central nervous system during the life history of the beetle, *Passalus cornutus* Fabricius. *Journal of Morphology*, **62**, 503–22.

Corneli, W. (1924). Von dem Aufbau des Sehorgans der Blattwespenlarven und der Entwicklung des Netzauges. *Zoologische Jahrbücher Abteilung für Anatomie und Ontogenie der Tiere*, **46**, 573–605.

Daly, H. V. (1964). Skeleto-muscular morphogenesis of the thorax and wings of the honey bee *Apis mellifica* (Hymenoptera: Apidae). *University of California Publications in Entomology*, **39**, 1–77.

Dietrich, W. (1909). Die Facettenaugen der Dipteren. *Zeitschrift für Wissenschaftliche Zoologie*, **92**, 465–539.

Edwards, J. S. (1969*a*). The composition of an insect sensory nerve, the cercal nerve of the house cricket *Acheta domesticus*. *27th Annual Proceedings of the Electron Microscope Society of America*.

(1969*b*). Postembryonic development and regeneration of the insect nervous system. In *Advances in Insect Physiology*, ed. J. W. L. Beament, J. E. Treherne & V. B. Wigglesworth, vol. 6, pp. 97–137. London & New York: Academic Press.

Gieryng, R. (1965). Veränderungen der histologischen Struktur des Gehirns von *Calliphora vomitoria* (L.) (Diptera) während der postembryonalen Entwicklung. *Zeitschrift für Wissenschaftliche Zoologie*, **171**, 80–96.

Günther, K. (1911). Die Sehorgane der Larve und Imago von *Dytiscus marginalis*. *Zeitschrift für Wissenschaftliche Zoologie*, **100**, 60–115.

Gymer, A. & Edwards, J. S. (1967). The development of the insect nervous system. I. An analysis of postembryonic growth in the terminal ganglion of *Acheta domesticus*. *Journal of Morphology*, **123**, 191–7.

Heywood, R. B. (1965). Changes occurring in the central nervous system of *Pieris brassicae* L. (Lepidoptera) during metamorphosis. *Journal of Insect Physiology*, **11**, 413–30.

Hinke, W. (1961). Das relative postembryonale Wachstum der Hirnteile von *Culex pipiens*, *Drosophila melanogaster* und *Drosophila*-Mutanten. *Zeitschrift für Morphologie und Ökologie der Tiere*, **50**, 81–118.

Ho, F. K. (1961). Optic organs of *Tribolium confusum* and *T. castaneum* and their usefulness in age determination (Coleoptera: Tenebrionidae). *Annals of the Entomological Society of America*, **54**, 921–5.

Jacobson, M. (1970). *Developmental Neurobiology*, pp. 1–465. New York: Holt, Rinehart & Winston.

Johansen, H. (1893). Die Entwicklung des Imagoauges von *Vanessa urticae* L. *Zoologische Jahrbücher Abteilung für Anatomie und Ontogenie der Tiere*, **6**, 445–80.

Johansson, A. S. (1957). The nervous system of the milkweed bug *Oncopeltus fasciatus* (Dallas) (Heteroptera, Lygaeidae). *Transactions of the American Entomological Society*, **83**, 119–83.

Lucht-Bertram, E. (1962). Das postembryonale Wachstum von Hirnteilen bei *Apis mellifica* L. und *Myrmeleon europaeus* L. *Zeitschrift für Morphologie und Ökologie der Tiere*, **50**, 543–75.

Menees, J. H. (1961). Changes in the morphology of the ventral nerve cord during the life history of *Amphimallon majalis* Razoumowski (Coleoptera, Scarabaeidae). *Annals of the Entomological Society of America*, **54**, 660–3.

Murray, F. V. & Tiegs, O. W. (1935). The metamorphosis of *Calandra oryzae*. *Quarterly Journal of Microscopical Science*, **77**, 405–95.

Neder, R. (1959). Allometrisches Wachstum von Hirnteilen bei drei verschieden grossen Schabenarten. *Zoologische Jahrbücher Abteilung für Anatomie und Ontogenie der Tiere*, **77**, 411–64.

Newport, G. (1832). On the nervous system of the *Sphinx ligustri*, Linn., and on the changes which it undergoes during a part of the metamorphosis of the insect. *Philosophical Transactions of the Royal Society of London*, B, **122**, 383–98.

(1834). On the nervous system of the *Sphinx ligustri*, Linn., (Part II) during the latter stages of its pupa and its imago state; and on the means by which its development is effected. *Philosophical Transactions of the Royal Society of London*, B, **124**, 389–423.

Nordlander, R. H. & Edwards, J. S. (1968). Morphological cell death in the post-embryonic development of the insect optic lobes. *Nature, London*, **218**, 780–1.

(1969*a*). Postembryonic brain development in the monarch butterfly, *Danaus plexippus plexippus*, L. I. Cellular events during brain morphogenesis. *Wilhelm Roux Archiv für Entwicklungsmechanik der Organismen*, **162**, 197–217.

(1969*b*). Postembryonic brain development in the monarch butterfly, *Danaus plexippus plexippus*, L. II. The optic lobes. *Wilhelm Roux Archiv für Entwicklungsmechanik der Organismen*, **163**, 197–220.

(1970). Postembryonic brain development in the monarch butterfly, *Danaus plexippus plexippus*, L. III. Morphogenesis of centers other than the optic lobes. *Wilhelm Roux Archiv für Entwicklungsmechanik der Organismen*, **164**, 247–60.

Nüesch, H. (1968). The role of the nervous system in insect morphogenesis and regeneration. *Annual Review of Entomology*, **13**, 27–44.

Osborne, M. P. (1966). Ultrastructural observations on adult and larval nerves of the blowfly. *Journal of Insect Physiology*, **12**, 501–7.

Panov, A. A. (1960). The structure of the insect brain during successive stages of postembryonic development. III. Optic lobes. *Entomological Review*, **39**, 55–68.

(1961). The structure of the insect brain at successive stages in post-embryonic development. IV. The olfactory center. *Entomological Review*, **40**, 140–5.

(1962). The nature of cell reproduction in the central nervous system of the nymph of the house cricket (*Gryllus domesticus* L., Orth., Insecta). *Doklady Akademii Nauk SSSR*, **143**, 471–4.

(1963). The origin and fate of neuroblasts, neurons and neuroglial cells in the central nervous system of the China oak silkworm *Antheraea pernyi* Guer. (Lepidoptera, Attacidae). *Entomological Review*, **42**, 186–91.

(1966). Correlations in the ontogenetic development of the central nervous system in the house cricket *Gryllus domesticus* L. and the mole cricket *Gryllotalpa gryllotalpa* L. (Orthoptera, Grylloidea). *Entomological Review*, **45**, 179–85.

Pipa, R. L. (1963). Studies on the hexapod nervous system. VI. Ventral nerve cord shortening; a metamorphic process in *Galleria mellonella* (L.) (Lepidoptera, Pyrallidae). *Biological Bulletin. Marine Biology Laboratory, Woods Hole*, **124**, 293–302.

(1967). Insect neurometamorphosis. III. Nerve cord shortening in a moth, *Galleria mellonella* (L.), may be accomplished by humoral potentiation of neuroglial motility. *Journal of Experimental Zoology*, **164**, 47–60.

(1969). Insect neurometamorphosis. IV. Effects of the brain and synthetic α-ecdysone upon interganglionic connective shortening in *Galleria mellonella* (L.) (Lepidoptera). *Journal of Experimental Zoology*, **170**, 181–92.

Pipa, R. L. & Woolever, P. S. (1964). Insect neurometamorphosis. I. Histological changes during ventral nerve cord shortening in *Galleria mellonella* (L.) (Lepidoptera). *Zeitschrift für Zellforschung*, **63**, 405–17.

Pipa, R. L. & Woolever, P. S. (1965). Insect neurometamorphosis. II. The fine structure of perineurial connective tissue, adipohemocytes, and the shortening ventral nerve cord of a moth, *Galleria mellonella* (L.). *Zeitschrift für Zellforschung*, **68**, 80–101.

Poulson, D. F. (1950). Histogenesis, organogenesis, and differentiation in the embryo of *Drosophila melanogaster* Meigen. In *Biology of Drosophila*, ed. M. Demerec, pp. 168–274. New York: Wiley.

Power, M. E. (1943). The effect of reduction in numbers of ommatidia upon the brain of *Drosophila melanogaster*. *Journal of Experimental Zoology*, **94**, 33–72.

(1946). An experimental study of the neurogenetic relationship between optic and antennal sensory areas in the brain of *Drosophila melanogaster*. *Journal of Experimental Zoology*, **103**, 429–62.

(1950). The central nervous system of winged but flightless *Drosophila melanogaster*. *Journal of Experimental Zoology*, **115**, 315–40.

Randall, W. C. (1968). Anatomical changes in the neuromuscular complex of the prolegs of *Galleria mellonella* (L.) (Lepidoptera: Pyralididae) during metamorphosis. *Journal of Morphology*, **125**, 105–27.

Randall, W. C. & Pipa, R. L. (1969). Ultrastructural and functional changes during metamorphosis of a proleg muscle and its innervation in *Galleria mellonella* (L.) (Lepidoptera: Pyralididae). *Journal of Morphology*, **128**, 171–94.

Richard, G. (1950). Le phototropisme du termite a cou jaune (*Calotermes flavicollis* Fabr.) et ses bases sensorielles. *Annals de Sciences Naturelles, Zoologique*, **12**, 485–507.

Robertson, J. & Pipa, R. L. (1973). Metamorphic shortening of interganglionic connectives of *Galleria mellonella* (Lepidoptera) *in vitro*: stimulation by ecdysone analogues. *Journal of Insect Physiology*, **19**, 673–9.

Sbrenna, G. (1971). Postembryonic growth of the ventral nerve cord in *Schistocerca gregaria* Forsk. (Orthoptera: Acrididae). *Bolletino di Zoologia*, **38**, 49–74.

Schaefer, P. E. (1938). The embryology of the central nervous system of *Phormia regina* Meigen (Diptera: Calliphoridae). *Annals of the Entomological Society of America*, **31**, 92–111.

Schrader, K. (1938). Untersuchungen über die Normalentwicklung des Gehirns und Gehirntransplantationen bei der Mehlmotte *Ephestia kuhniella* Zeller nebst einigen Bemerkungen über das Corpus allatum. *Biologisch Zentralblatt*, **58**, 52–90.

Schwager-Hübner, M. (1970). Untersuchungen über die Entwicklung des thorakalen Nerven–Muskel-Systems bei *Apis mellifica* L. (Hymenoptera). *Revue Suisse de Zoologie*, **77**, 807–49.

Scudder, G. G. E. & Hewson, R. J. (1971). The postembryonic development of the indirect flight muscles in *Oncopeltus fasciatus* (Dallas) (Hemiptera: Lygaeidae). *Canadian Journal of Zoology*, **49**, 1377–86.

Snodgrass, R. E. (1954). Insect metamorphosis. *Smithsonian Miscellaneous Collections, Washington, DC*, **122** (9), 1–124.

Springer, C. A. (1967). Embryology of the thoracic and abdominal ganglia of the large milkweed bug, *Oncopeltus fasciatus* (Dallas), (Hemiptera, Lygaeidae). *Journal of Morphology*, **122**, 1–18.

Teutsch-Felber, D. (1970). Experimentelle und histologische Untersuchungen an der Thoraxmuskulatur von *Periplaneta americana* L. *Revue Suisse de Zoologie*, **77**, 481–523.

Tiegs, O. W. (1922). Researches on the insect metamorphosis. Part 1. On the structure and post-embryonic development of a chalcid wasp, *Nasonia*. *Transactions of the Royal Society of South Australia*, **46**, 319–491.

Tung, A. S.-C. & Pipa, R. L. (1971). Fine structure of transected interganglionic connectives and degenerating axons of wax moth larvae. *Journal of Ultrastructure Research*, **36**, 694–707.

(1972). Insect neurometamorphosis. V. Fine structure of axons and neuroglia in the transforming interganglionic connectives of *Galleria mellonella* (L.) (Lepidoptera). *Journal of Ultrastructure Research*, **39**, 556–67.

Wigglesworth, V. B. (1956). Formation and involution of striated muscle fibres during the growth and moulting cycles of *Rhodnius prolixus* (Hemiptera). *Quarterly Journal of Microscopical Science*, **97**, 465–80.

Woolley, T. A. (1943). The metamorphosis of the nervous system of *Aedes dorsalis* Meigen (Diptera: Culicidae). *Annals of the Entomological Society of America*, **36**, 432–47.

Neural specificity as a game of cricket: some rules for sensory regeneration in *Acheta domesticus*

J. S. Edwards and J. Palka

Introduction

The ability of arthropods to regenerate lost parts has been known for centuries. From the large literature that has accumulated over the years several generalisations emerge. We can say that regeneration of appendages occurs only in immature animals, or in adults induced to undergo an extra moult, and that the process of limb regeneration is tied by feedback mechanisms to the endocrine timing of the moulting cycle. We also know that appendages show axial gradients in their capacity for regeneration and that heteromorphic or multiple regenerates may be induced predictably in a wide range of arthropods. Regenerates arise from specific areas encompassing a so-called organ field surrounding the appendage. Normally they repeat the ontogeny of the original appendage in a condensed sequence to produce a functional replacement which, under optimal conditions, can be structurally almost identical with the original limb. While regenerates normally resemble the original limb, many striking examples of heteromorphic regenerates are known (Needham, 1965), some of which are of considerable neurological interest, for example the leg-like forms which develop from excised antennae of the phasmids, *Carausius* (Brecher, 1924) and *Sipyloidea* (Urvoy, 1970) or the antennules that develop from excised eye stalks in lobsters (Herbst, 1896; Maynard, 1965).

Trophic role of nerves in the regeneration of insect appendages

The focus of interest in regeneration studies with insects has generally been on the development of the integument, and although there are long-standing generalisations on the role of the nervous system, many questions remain open. The earliest of these stems from the reports of Kopéc (1923), Friedrich (1930) and Suster (1933) that appendages could regenerate in the absence of accompanying ganglia, but that the regenerates so formed lacked functional musculature. The ectodermal parts of an appendage can evidently initiate regeneration autonomously, while the mesodermal parts require innervation (Urvoy, 1959). But the removal of a ganglion does not necessarily deprive an appendage of innervation, for as Kopéc (1923) and Bodenstein (1955, 1957) demonstrated, vigorous sprouting occurs from cut connectives. This finding, that it is virtually impossible to eliminate with certainty all sources of re-innervation, places a restraint on the interpretation of nerve extirpation experiments.

The vigour and rapidity with which cut motor nerves of arthropods can develop outgrowths is not generally recognised and claims for the necessity of innervation in regeneration must be taken with some caution, particularly since it is now clear that invasion of deprived areas, not necessarily by the functionally correct motor neurons, seems to be a general phenomenon. Yet, limb regenerates transplanted to a dorsal position do not thrive (Bodenstein, 1953; Steinberg, 1959), so perhaps segmental gradient effects (Bate & Lawrence, this volume) or nerve dosage effects (Singer, 1965) are also involved.

While there is no doubt that motor innervation is essential for the full development of muscle in insects, as Nüesch (1968) and others have shown, this trophic relationship, which is shared with vertebrates, is not so clear in the case of sensory appendages which lack intrinsic musculature. Drescher's (1960) studies of antennal regeneration after brain ablation in *Periplaneta americana* do imply the necessity for central connections in the regeneration of a sensory structure, but in other cases confusion has arisen from the direct extrapolation of the well known vertebrate findings to insects. Rummel (1970), for example, argues on the basis of surgical and radiation procedures that the regeneration of abdominal cerci in the house cricket *Acheta domesticus* is dependent on centrifugal innervation in a manner comparable to limb regeneration in vertebrates. But, as with Schoeller's (1964) claim for the necessity of motor innervation for the differentiation of imaginal discs in *Sarcophaga*, it is necessary to determine whether observed nerve connections are composed of centrifugal fibres, or whether they arise solely from sensory neurons that differentiate in the integument and reach the CNS.

Just as the differing architecture of vertebrate and arthropod nervous systems makes direct comparisons of the role of sensory and motor fibres in regeneration hazardous, it is also problematic how far nerve section experiments in arthropods can be compared with those of vertebrates. In most cases the approach for nerve section at the base of an appendage disrupts circulation and tracheation. Thus necrosis observed in appendages after nerve section (e.g. Rummel, 1970) is likely to be the result of nutritional and respiratory failure rather than of trophic relationships requiring intact innervation. Cercal regenerates complete with sensilla will differentiate after major damage to the terminal ganglion designed to disrupt innervation, (McLean & Edwards, unpublished results) and in our estimation the necessity for neural connections in the differentiation of a sensory appendage remains an open question.

Sensory regeneration and neural specificity

Despite extensive studies on the phenomenology of regeneration, recognition of the potential of arthropods for the study of neural specificity is quite recent and follows by many years the pioneer studies, all of which were based on vertebrates and are amply reviewed in recent books mentioned in the editor's introduction to this volume.

The arthropods are in many ways pre-adapted to studies in neural development and regeneration. To cite a few special characteristics, we can note that:

1. The architecture of the nervous system is convenient, with motor cell bodies and interneurons located centrally and sensory cells in the integument where they serve both to transduce and to transmit to the centre, and where they can be removed by simple surgery so that neural regeneration occurs *de novo*.

2. Postembryonic growth and differentiation proceeds through a series of stages, each marked by a cycle of growth and moulting under endocrine control, and subject to experimental intervention.

3. Sensory systems grow during postembryonic development by the cyclic addition of new neurons which differentiate in the integument and augment sensory projection to the central ganglia during each instar. They terminate on a central population of neurons which remains constant in number throughout postembryonic development.

4. Arthropods regenerate appendages during postembryonic development, and in doing so, recapitulate the ontogeny of the appendage. They survive outrageous surgical insults and accept grafts from their own and sometimes other species with relative ease.

5. Their small size and parsimonious neuron populations make analyses at the cell-to-cell level possible in physiological and ultrastructural studies.

Systems with these attributes can reveal much about the social behaviour of neurons, for example the response of the developing neuron to its environment, and the degree of rigour surrounding the generation of connections. Cell surface phenomena undoubtedly underlie the crucial mechanisms that establish final connectivity, and it seems doubtful that the mechanism could be elucidated at the molecular level in the thicket of the natural neuropile. But the rule book must first be constructed, and that is the goal of our studies on regeneration in insects.

Regeneration of cricket cerci

As an example of the potential of arthropods for studies in neural specificity we will present a summary of our work on the regeneration of abdominal cerci in the house cricket, *Acheta domesticus*. The point of departure for these studies was the demonstration that regenerating abdominal cerci could be transplanted to the thorax and establish functional connections with the central nervous system (Edwards & Sahota, 1967). This study was restricted by the limitations of light microscopy for fine insect neuroanatomy, and by our rudimentary knowledge of the organisation of the giant interneurons in the thoracic region. The question is now being approached at a level that can be expected to give answers at the level of cellular interactions. Our first task is to characterise the behaviour of neurons when the abdominal cercus regenerates *in situ* and we shall outline our results below.

Normal cerci

The essential features of the anatomy of the abdominal cerci and their neural connections are summarised below. The cerci are paired sensory appendages of the 11th abdominal segment, and are

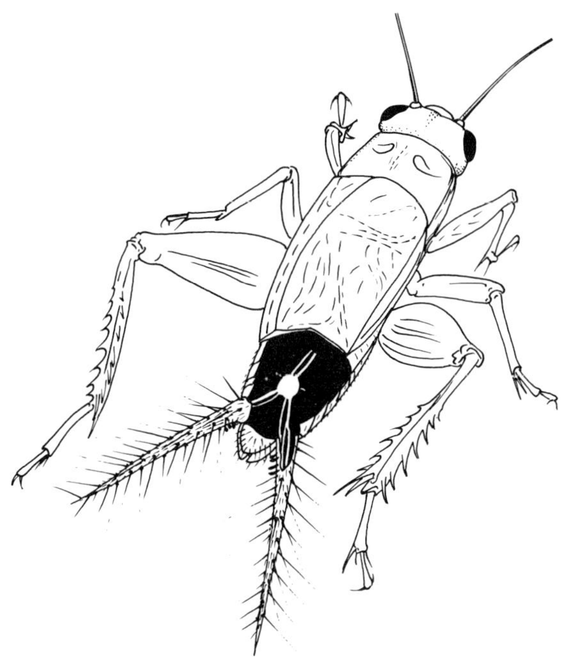

Fig. 1. Adult male house cricket, *Acheta domesticus*. Wings and dorsum of posterior abdomen removed to show terminal ganglion which has been raised from normal position to demonstrate cercal sensory nerves entering cerci, and cercal motor nerves which supply the extrinsic cercal musculature. Other nerves associated with the terminal ganglion are not shown.

characteristic of orthopteroid insects, e.g. crickets, grasshoppers and cockroaches. They are exceptionally well-developed in crickets (Gryllidae) (Fig. 1). They take the form of elongate cones, densely clothed with sensory hairs (Plate 1*a*) and other mechanoreceptors whose cell bodies lie in the epidermal layer underlying the cuticle from which the sensory hairs arise. Other than tracheae and haemolymph passages bounded by connective tissue, these are the only structural elements of the cercus. A set of extrinsic muscles inserts externally at the base of the cercus. Five receptor types are present on the cerci of *Acheta* of which three are varieties of mobile mechanoreceptor hairs or sensilla trichodea. The most conspicuous of these sensilla are about 150 long, mobile hairs, among which are interspersed about 150 smaller, mobile hairs. A group of about 65 clavate, mobile hairs are situated on the mesial

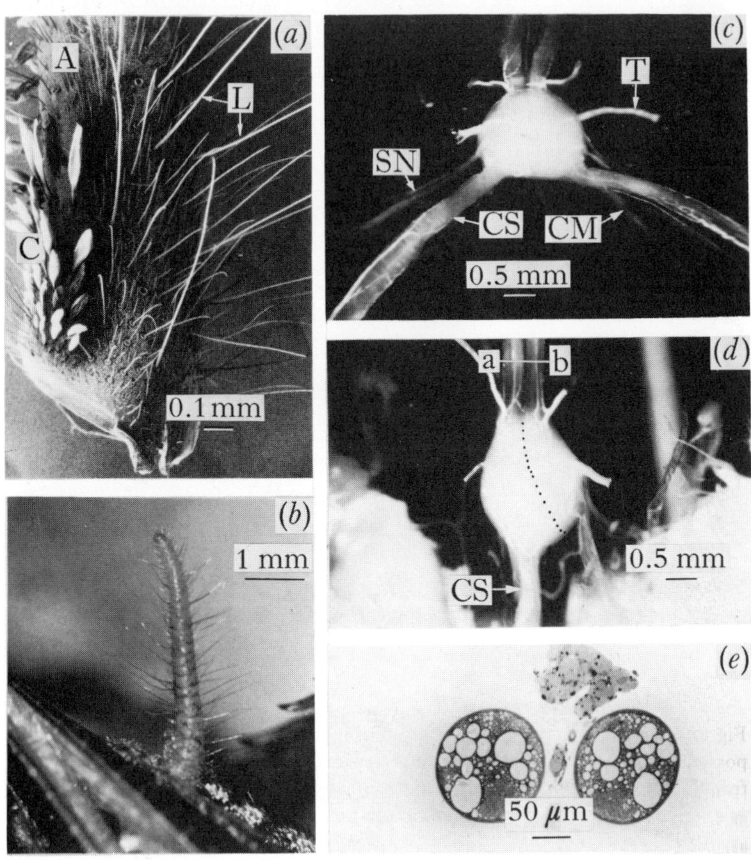

Plate 1. (*a*) Scanning electron micrograph of base of adult cercus. The longest of the sensory hairs, the vibrissa-like hairs (L) arise from a mobile articulation within a cuticular socket. Clavate hairs (C) on the mesial surface are similarly articulated. Numerous appressed sensilla chaetica (A) cover the entire cercus. The cuticular surface of the cercus is clothed with minute non-sensory scales.

(*b*) Regenerate cercus on adult, formed during the last three larval instars. All types of sensilla are present.

(*c*) Normal adult terminal ganglion and some associated nerves. CM, cercal motor nerve; CS, cercal sensory nerve; SN, segmental nerves; T, trachea.

(*d*) Terminal ganglion of adult which developed with right cercus absent throughout postembryonic development. Labelling as for (*c*). In the absence of the right cercal sensory nerve, the right half of the ganglion is markedly atrophied. Dotted line marks the anatomical midline.

(*e*) Transverse section (ab) of ventral nerve cord shown in (*d*). Despite marked asymmetry of ganglion, development of the giant interneurons is nearly symmetrical.

base of the cercus. About 2200 multiply-innervated trichoid sensilla, set deeply in appressed sockets may function as chemoreceptors as well as mechanoreceptors, though this remains to be proved. The fifth category of cercal receptor includes campaniform sensilla situated at the base of sockets surrounding mobile hairs. They presumably function as cuticular strain gauges which monitor the deformation of hair sockets under strong stimulation.

Sensory axons arising from cercal receptors assemble in groups which coalesce to form two bundles in the base of the cercus, finally leaving the cercus as a single, stout, purely sensory nerve which enters the terminal abdominal ganglion at its posterior–lateral angle (Plate 1*c*). The cercal motor nerve supplying the cercal musculature runs parallel to the sensory nerve, but never fuses with it. The cercal nerve thus carries only sensory fibres from the cercus, together with a small population of sensory fibres from sensory hairs on periproct and paraproct plates which form the terminal sensilla of the abdomen.

The cercal sensory nerve is a relatively stout structure bearing about 10000 axons (Edwards, 1969) ranging in diameter between 7 and 0.05 μm. The number of dendrites associated with the receptors enumerated above approximates to the number of cercal nerve axons; there appears to be neither fusion nor collateral formation in the cercal nerve.

Fibres from the cercal nerve terminate in the last abdominal ganglion. It has not been established that all cercal sensory fibres terminate there, but it is clear that the majority of fibres do project to two principal areas of the terminal ganglion (Fig. 2). The major projection of the cercal nerve lies principally in the dorsal part of the posterior ipsilateral quadrant in a loosely glomerular structure. Terminations also occur more toward the median plane in the mid region of the ganglion. A second glomerular region lies anteroventrally. Terminations of the cercal nerve are associated with collaterals of the giant fibre system to which they evidently project directly. No contralateral terminations have been detected in degeneration studies of normal animals.

Electrophysiological recording from the surface of the abdominal connectives of *Acheta domesticus* immediately reveals substantial ongoing spike activity, in the range of 0.8–1.5 mV amplitude and 15–25 s^{-1} frequency (Plate 2). A mechanical stimulus, such as a tone or an air puff, may accelerate or inhibit this activity depending

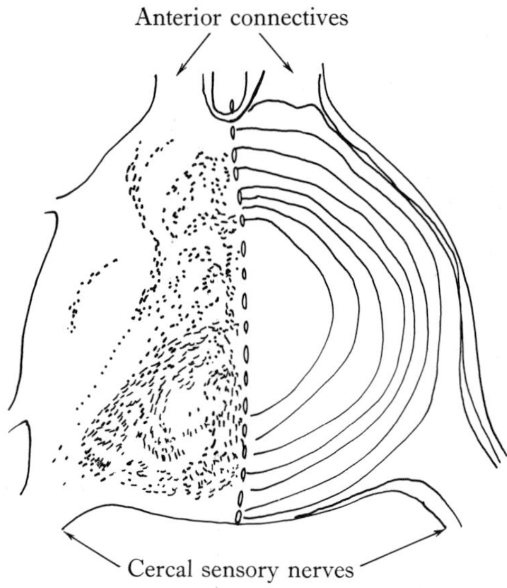

Fig. 2. Schematic diagram based on serial sections of terminal ganglion showing partial representation of degeneration sites following amputation of cercus. Degeneration sites are represented by small, black dots on the left. Small ellipses mark midline and contours of dorsal surface of ganglion are shown on the right. Degeneration is entirely ipsilateral after removal of one cercus from a normal animal.

on such of its characteristics as intensity, frequency, and direction. The response properties of these units are complex and only partially understood.

The largest and best studied units in the cord are normally silent, but can be excited by air puffs, sound and substrate vibration. Confirmation that the largest spikes seen in multiunit recordings are produced by the largest axons seen in cross sections of the connectives comes from the recent work of Murphey (in press) who has made simultaneous intra- and extracellular recordings and marked the cells with Procion Yellow.

For the purpose of testing regenerate animals, we have measured the response of the giant fibres to tones in the frequency range of 40–2000 Hz, to air puffs whose peak velocities range from subthreshold to saturation of the response, and to substrate vibration produced by dropping a solenoid plunger from various heights, this stimulus also ranging in intensity from subthreshold to saturation.

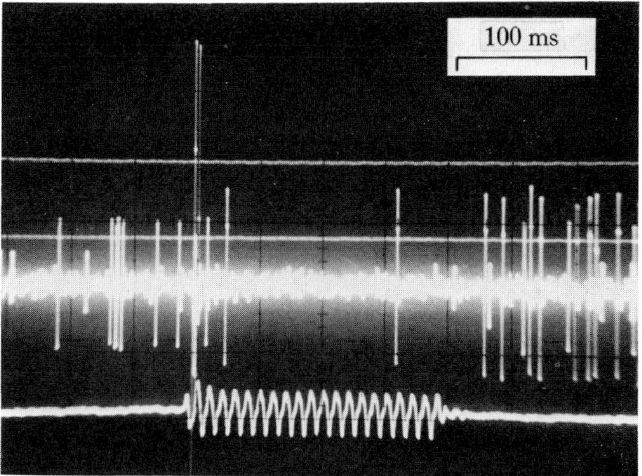

Plate 2. Multiunit, extracellular recording from the abdominal ventral nerve cord of a normal adult, showing response to a tone pulse of 100 Hz at 85 dB. Two spikes from the largest giants occur at the onset of the tone; the irregular, tonic firing of smaller giants is inhibited. The two marker traces monitor the upper and lower threshold levels of the window circuit used to select spike sizes for quantitative study.

The spikes of these largest giants are 2–5 mV in amplitude. There are probably two units involved, and in many preparations their spikes are readily distinguishable, but we have studied them together in the present experiments. On those occasions when the units have been examined individually, their responses to our standard stimuli have been identical. Some of the known response properties of these cells will become apparent in the descriptions of regenerate animals which follow.

Patterns of cercal regeneration

Cerci regenerate vigorously (Plate 1*b*). The pattern of regeneration depends on culture conditions; our crickets, which are raised under near optimal conditions, pass through nine immature instars before moulting to the adult and require five to six instars to regenerate a fully symmetrical cercus. The detailed pattern depends on time of excision of the cercus, and it will suffice here to outline the events following loss of a cercus immediately after a moult. Early instars so treated produce a small fragile button after the first moult which very rarely has sensilla on its surface. The second moult following onset of regeneration yields a small cercus clothed with sensilla of all the types outlined above. Neither removal of cerci immediately after the moult, nor subsequent regeneration affects instar length, and no significant differences in other body dimensions are incurred following regeneration.

Crickets deprived of cerci at the time of hatching, and whose regenerates are removed at each successive moult develop without cercal sensory input to the terminal ganglion. Thus in such animals the only known contacts formed between cercal sensory axons and abdominal giant fibres are those formed during embryonic development, which may function for a matter of minutes at the time of hatching and before the cercus is removed.

Pioneer cercal sensilla of the first instar may of course play a significant role in the differentiation of the terminal ganglion, but it should be noted that the sensilla are few in number (about 65) and send about 240 axons to the ganglion, while the adult's 2800 hairs and hair-associated campaniform sensilla give rise to about 10000 axons. Animals raised with one cercus throughout life only incur a slight delay in development, and otherwise appear normal. It is possible by means of successive extirpations to raise crickets deprived of all cercal input, or of input to one side only. Since the operation of removing the cercus at its base does not also remove the regenerative tissue surrounding the cercus, it is possible to delay development of a cercus until later instars. As is characteristic of regenerates, the delayed cercus recapitulates ontogenetic stages during the development of its sensillar population. Sensory cells can differentiate within one instar in a regenerate cercus, and neuron profiles have been found in sections of early regenerates, but the timing of their arrival centrally remains to be determined.

We shall now consider several patterns of cercal regeneration:

1. *Both cerci removed up to the 6th instar, and both regenerating from the 7th to the adult stage.* Cerci with all categories of sensilla are regenerated, and stout cercal sensory nerves make contact with the terminal ganglion. Neuropile volume is reduced in the terminal ganglion, the loss being principally in the posterior lateral angles.

Giant fibre responses in both connectives are normal in the following respects: the fibres show a very low level of spiking in the absence of deliberate stimulation of the cerci; they respond to tones and have the same tuning curve as in normal animals; they respond to air puffs; they respond only very poorly to substrate vibration, and their excitatory inputs are ipsilateral.

2. *Both cerci removed up to the 6th instar; one cercus permitted to regenerate from the 7th to the adult stage.* In such animals the regenerate cercus has all sensilla, a cercal sensory nerve enters the ganglion, and ganglion neuropile is asymmetric. A small population of contralaterally projecting cercal fibres are found in degeneration preparations.

Giant fibre activity in the connective ipsilateral to the regenerate cercus is normal in response to auditory and air puff stimulation, but slightly enhanced in response to substrate vibration. The vibration response is yet more pronounced on the chronically deprived side. It persists in both connectives when the regenerate cercus is removed, and must therefore be the result of input from non-cercal receptor elements.

There is, in these animals, a response of the giant fibres to air puffs on the side lacking a cercus. It is small but quite reliable, even at low air velocities. For example, ipsilateral responses in normal and unilateral regenerate animals averaged 10.40 and 9.14 spikes per response respectively to one of our standard stimuli; the responses on the contralateral side averaged 0.07 spikes in normal animals and 1.42 spikes in experimentals. The contralateral responses, therefore, were about 20 times as great in experimentals as in normals. They disappeared when the single, regenerate cercus was removed.

3. *Growth of one cercus prevented throughout development to produce an asymmetrical animal with a single cercus.* In this case the growth of the terminal ganglion undergoes major modification. The neuropile volume on the operated side is reduced by about 30% and cortical volume is reduced by a similar proportion (Plate 1*d*). Despite this diminution in neuropile volume, the giant axons, at least where they enter the abdominal connectives, are not significantly different in cross sectional area (Plate 1*e*). Further, intracellular dye injection indicates that the major branches of at least one giant fibre are normal. We can infer that the giants are not dependent in a quantitative manner on their peripheral contacts.

Degeneration procedures reveal the presence of a small but distinct population of fibres that have crossed the midline of the ganglion, a situation not encountered in normal animals.

The results of physiological study of giant fibre responses have been similar to those just described for animals which had regenerated a single cercus, but even more pronounced. Auditory responses are purely ipsilateral, and their frequency dependence is as for normal animals. Contralateral responses to puff stimuli are even greater, averaging 2.75 spikes per response compared with the 0.07 spike average of control animals (the ipsilateral responses were 10.50 and 10.40 respectively). The response to substrate vibration is normal on the intact side, but six times greater than normal on the deprived side. The contralateral puff response is dependent on the single cercus, whereas the vibration response is not.

4. *Growth of cercus on one side prevented for 6 instars, development then reversed with removal of intact cercus and development of delayed regenerate.* In this case a vastly increased incidence of contralateral degeneration is observed. The simultaneous degeneration of cercal fibres on one side of the ganglion, and regeneration on the other appears to *maximise* the occurrence of contralateral fibres.

These animals show many anomalies in their electrophysiological responses, and will not be considered here.

Discussion

The major results of this study are the following:

1. Normal connections with the giant fibre system can be re-established after a long delay, comprising at least two-thirds of postembryonic development. This raises the question of the pattern of growth of receptor sites on giant interneurons: we know that the number of sensory fibres projecting to the terminal ganglion increases more than 40-fold during normal postembryonic development, and we must assume that the sensory fibres which reach interneuron collaterals in later instars from newly added receptors must find sites upon the existing interneurons, for the number of neurons in the ganglion does not increase during postembryonic development (Gymer & Edwards, 1967). Perhaps

the most instructive way to view these experiments is that we are observing a regenerating sensory system projecting to a growing CNS.

2. The regenerative process comes close to restoring normal patterns of inputs to the giant fibres when regeneration is symmetric. Where postembryonic development of the cerci is asymmetric as a result of cercus amputation and removal of successive regenerates, errors can be detected, both in the anatomical projection of primary afferents and in the physiological responses monitored in giant interneurons. Fibres cross the midline in a manner not seen in normal development, and contralateral excitation of giant fibres by cercal sensilla responsive to air puffs, and by non-cercal receptors excited by substrate vibration, occurs in a manner not seen in symmetrical animals.

3. Where normal cercal input is absent, other inputs can terminate on giant interneurons. Thus, the vibration receptors, which have been recognised physiologically but not located anatomically, provide greatly enhanced input to the giant fibres in the absence of normal cercal input. Pioneer fibres from these receptors might be augmented during subsequent instars as a result of contact guidance (Wigglesworth, 1953).

The competence of neuronal interaction late in postembryonic development in the insect seems to contrast with the relatively short period available for compensatory developmental changes in the mammalian nervous system. But direct comparisons are perhaps inappropriate since we are dealing in the insect with a system that, even in normal individuals, is continuing to add sensory elements to the central neuropile throughout postembryonic development. Perhaps a closer comparison would be afforded by studying the connections of grafted cerci in adult crickets induced to moult by means of ecdysone injection.

We shall not explore in detail here the comparisons with vertebrate regeneration, but some, which arise during asymmetric development of sensory systems and which may tell us something about neuronal plasticity, should be noted. For example, fibres of the optic tract of rats follow altered routes in animals deprived of one eye early in postnatal life. It is now known from several studies of the mammalian visual system that an afferent fibre population may change its projection in response to the absence of other inputs: partial denervation leads to an increased projection by the remaining afferents (Liu & Chambers, 1958; Rose, Malis, Kruger & Baker, 1960; Goodman & Horel, 1966; Raisman, 1969).

The mechanisms by which this may occur are beyond our present reach. Liu & Chambers suggested that guidance to growing fibres may be provided by the degenerating terminals or their glia. Lund & Lund (1971) and Raisman (1969) have recently shown that postsynaptic sites vacated by section of the presynaptic axons persist, and tend to become occupied in a rather non-specific way by processes from other available neurons. Thus, the postsynaptic elements may themselves be foci of attraction for regenerating axons.

In our experimental material we have seen both redistribution of axons of normal cercal origin when development is asymmetric, and the enhancement of non-cercal input. The fact that maximal crossing of cercal axons is seen when regeneration on one side coincides with massive degeneration on the other might be expected if guidance of regenerating fibres emanates from degenerating terminals or their glia. But this does not account for contralateral connections which are formed in animals which develop with one cercus only, unless it is the degeneration of the small population of contacts formed within the egg and disrupted at hatching. It may be that fibres which are drawn across the midline at this time function as pioneer fibres which serve as a guide to a small population of contralateral fibres during subsequent development.

We know that all of the neurons intrinsic to the terminal ganglion are present at hatching (Gymer & Edwards, 1967), and grow in volume even in the absence of cercal input (Edwards & Palka, 1971). Do the giant fibres develop postsynaptic sites in anticipation of each new instar's crop of projecting cercal axons? If so, their failure to become occupied by the 'intended' afferent fibres might make them attract other available afferents, as suggested in vertebrate material by Raisman and by the Lunds.

Our object in the work described above has been to try to define the canons of neurogenesis as they apply to the regenerating cercal sensory system. Thus far we have been able to show that neurons arising *de novo* in regenerates can project normally to the ganglion and that the quality of repair is not impaired by prolonged absence of contact, but that asymmetric development does give rise to errors in sidedness and that the input of one modality may be augmented by the reduction of another in seemingly competitive fashion.

What else can we hope to learn? We can ask if unusual connections induced by cercal retardation are correctable. And now that individual giant interneurons can be identified and labelled, we can begin to ask questions about the behaviour of growing or regenerating afferent fibres with respect to particular cells. How, for example, does each new batch of cercal sensilla insert on the interneuron: do they contact in a mosaic or in growth zones? Answers to such questions should allow us to write further parts of the rule book for specificity and plasticity in the nervous system.

References

Bodenstein, D. (1953). Regeneration. In *Insect Physiology*, ed. K. D. Roeder, 866–78. New York: Wiley.

(1955). Contributions to the problem of regeneration in insects. *Journal of Experimental Zoology*, **129**, 209–24.

(1957). Studies on nerve regeneration in *Periplaneta americana*. *Journal of Experimental Zoology*, **136**, 89–115.

Brecher, L. (1924). Die Bedingungen für Fühlerfüsse bei *Dixippus* (*Carausius*) *morosus* Br. et Redt. *Archiv für Entwicklungsmechanik der Organismen*, **102**, 549–72.

Drescher, W. (1960). Regenerationsversuche am Gehirn von *Periplaneta americana* unter Berücksichtigung von Verhaltensänderung und Neurosekretion. *Zeitschrift für Morphologie und Ökologie der Tiere*, **48**, 576–649.

Edwards, J. S. (1969). Composition of an insect sensory nerve, the cercal nerve of the house cricket *Acheta domesticus*. *Proceedings of the Electron Microscopical Society of America*, **17**, 248–9.

Edwards, J. S. & Palka, J. (1971). Neural regeneration: delayed formation of central contacts by insect sensory cells. *Science*, **172**, 591–4.

Edwards, J. S. & Sahota, T. S. (1967). Regeneration of a sensory system: the formation of central connections by normal and transplanted cerci of the house cricket *Acheta domesticus*. *Journal of Experimental Zoology*, **166**, 387–96.

Friedrich, W. H. (1930). Zur Kenntnis der Regeneration der Extremitäten bei *Carausius morosus*. *Zeitschrift für Wissenschaftliche Zoologie*, **137**, 578–605.

Goodman, D. C. & Horel, J. A. (1966). Sprouting of optic tract projections in the brain stem of the rat. *Journal of Comparative Neurology*, **127**, 71–88.

Gymer, A. & Edwards, J. S. (1967). The development of the insect nervous system. I. An analysis of postembryonic growth in the terminal ganglion of *Acheta domesticus*. *Journal of Morphology*, **123**, 191–7.

Herbst, C. (1896). Über die Regeneration von antennenähnlichen Organen an Stelle von Augen. VII. Die Anatomie der Gehirnnerven und des Gehirnes bei Krebsen mit Antennulis an Stelle von Augen. *Archiv für Entwicklungsmechanik der Organismen*, **42**, 407–89.

Kopéc, S. (1923). The influence of the nervous system on the development and regeneration of muscles and integument in insects. *Journal of Experimental Zoology*, **37**, 15–25.

Lui, C. N. & Chambers, W. W. (1958). Intraspinal sprouting of dorsal root axons. *Archives of Neurology and Psychiatry*, **79**, 46–61.

Lund, R. D. & Lund, J. S. (1971). Synaptic adjustment after deafferentation of the superior colliculus of the rat. *Science*, **171**, 804–7.

Maynard, D. M. (1965). The occurrence and functional characteristics of heteromorph antennules in an experimental population of spiny lobsters *Panulirus argus*. *Journal of Experimental Biology*, **43**, 79–106.

Needham, E. A. (1965). Regeneration in the Arthropoda and its endocrine control. In *Regeneration in Animals and Related Problems*, ed. V. Kiortsis & H. A. L. Trampusch, pp. 283–323. Amsterdam: North Holland Publishing Co.

Nüesch, H. (1968). The role of the nervous system in insect morphogenesis and regeneration. *Annual Review of Entomology*, **13**, 27–44.

Raisman, G. (1969). Neuronal plasticity in the spetal nuclei of the adult rat. *Brain Research*, **14**, 25–48.

Rose, J. E., Malis, L. I., Kruger, L. & Baker, C. P. (1960). Effects of heavy, ionizing, monoenergetic particles on the cerebral cortex. II. Histological appearance of laminar lesions and growth of nerve fibre after laminar destruction. *Journal of Comparative Neurology*, **115**, 243–95.

Rummel, H. (1970). Die nerveninduzierte Regeneration der Cerci bei *Acheta domesticus* L. *Deutsche Entomologische Zeitschrift*, **17**, 357–409.

Schoeller, J. (1964). Recherches descriptives et expérimentales sur la céphalogenèse de *Calliphora erythrocephala* Meig. *Archives de Zoologie expérimental et générale*, **103**, 1–216.

Singer, M. (1965). A theory of the trophic nervous control of amphibian limb regeneration, including a re-evaluation of quantitative nerve requirements. In *Regeneration in Animals and Related Problems*, ed. V. Kiortsis & H. A. L. Trampusch. Amsterdam: North Holland Publishing Co.

Steinberg, D. M. (1959). Regeneration in homografted and heterografted limbs in the stick insect. *Doklady Akademii Nauk SSSR*, **129**, 1001–3 (Translation).

Suster, P. M. (1933). Vorderbeinregeneration nach Ganglion extirpation bei *Dixippus morosus*. *Anzeiger Akademic der Wissenschaften, Wien*, **70**, 65–6.

Urvoy, J. (1959). Étude de la régénération des organes sensoriels antennaires chez *Carausius* (*Dixippus*) *morosus*, Brunner. *Annales des sciences Naturellas Zoologie*, **25**, 309–16.

(1970). Etude des phénomènes de régénération après section d'antenne chez le Phasme *Sipyloidea sipylus* W. *Journal of Embryology and Experimental Morphology*, **23**, 719–28.

Wigglesworth, V. B. (1953). The origin of sensory neurones in an insect, *Rhodnius prolixus* (Hemiptera). *Quarterly Journal of Microscopical Science*, **94**, 93–112.

Postembryonic development of insect motor systems

D. R. Bentley

Introduction

In the field of arthropod neurobiology, most of the scientific effort has been directed toward understanding fully constructed nervous systems, with the view that it would be easier to analyse a stable system than one undergoing change. Recently, increasing attention has been given to how nervous systems come into being, including their development, genetics, and evolution. Developmental investigations traditionally have concentrated on neuroanatomy, and upon the embryonic period. In this paper, current research on the physiology and structure of motor neuronal networks during postembryonic development will be reviewed.

The goals of this research are: (i) to describe the sequence of physiological and structural events during development, (ii) to determine the chain of causality underlying these events, hopefully resulting in useful generalisations, (iii) to establish a link with genetics, and (iv) to illuminate unresolved problems with adult nervous systems. A promising approach to these goals is the method, so successful with adult systems, of investigating individually identified neurons which can be firmly associated either with sensory input or with behavioural output. Dealing with the same cell in repeated preparations has proved to be a remarkably powerful analytical tool.

In adopting this approach to developmental studies of motor systems, several characteristics are desirable. First, if the behaviour produced by the mature system is highly stereotyped, it provides a standard pattern against which the output of the immature network can be evaluated. Second, the behaviour should be generated by a reasonably small number of neurons identifiable by their peripheral connections or central properties. Third, if maturation occurs over an extended period, it is easier to discriminate stages of development, and, if the period is postembryonic, restrictions on experiments imposed by small size are less severe. These criteria are well met by the development of adult behaviour patterns, such as flight, in Orthoptera or other large, non-metamorphosing insects. In this review, attention will be confined almost exclusively to these preparations.

Accessory information

Although developmental neurophysiology of insects is a relatively new subject, valuable information is available in related areas. In this section, the types of information and some pertinent conclusions are described. First, the study of development is bracketed by extensive research on insect genetics and on the neurobiology of adult animals. The neuronal basis of flight, locomotion, and similar behaviour has been reviewed recently by Wilson (1968) and Hoyle (1970). In insect neurogenetics, recent work has dealt with the effect of single genes on the nervous system (Ikeda & Kaplan, 1970) and with the genetics of neuronal circuits generating stereotyped behaviour (Bentley, 1971; Bentley & Hoy, 1972). Some of the work with single loci has been summarised by Benzer (1971).

The study of postembryonic development is supported by numerous detailed examinations of insect embryology. About crickets alone, for example, descriptions are available for species in several genera, including *Teleogryllus commodus* (Brookes, 1952), *Acheta domesticus* (Kanellis, 1952; Mahr, 1961; Jobin & Huot, 1966; Heinig, 1967), *Gryllus assimilis* (Rakshpal, 1962), and *Scapsipedus marginatus* (Grellet, 1961). Analysis of orthopteran embryology is also available through tissue culture of crickets (Grellet, 1968), locusts (Mueller, 1963) and cockroaches (Levi-Montalcini, this volume), and through experiments *in vivo* with identified components of neuro-muscular systems (Tyrer, 1969).

Within the postembryonic period, descriptions of neuro-anatomy, musculature, and behaviour are particularly useful. Clearly, hatchling nymphs execute many behaviour patterns, including locomotion, feeding, and grooming, and must have an appropriate complement of neuro-muscular machinery. Among non-metamorphosing (hemimetabolous) insects, however, behavioural information on nymphs is rather meagre, with the exception of moulting related activities. Bernays (1971, 1972*a*, *b*) has provided an excellent study of locust hatchling behaviour up to the first ecdysis, with supporting accounts of musculature and its use. An important fact emerging from such analysis is that in hemimetabolous insects, nymphal behaviour is not merely an impoverished adult repertoire. Nymphs perform a variety of specialised behaviour patterns, sometimes only once in the life cycle, for which integrative and mechanical apparatus must be constructed and subsequently dismantled. For example, within five minutes of reaching the surface, *Schistocerca* nymphs will no longer react to burial by digging. These behaviour patterns should provide many opportunities for examining the construction of neuronal circuits generating particular patterns.

Development of musculature

The muscles of hemimetabolous insects grow through various combinations of direct enlargement, fibre cleavage, and myoblast incorporation (Tiegs, 1955; Hinton, 1959). Descriptions of larval musculature and its fate are available for several orthopteran types (Voss, 1911, 1912; Wiesend, 1957; Teutsch-Felber, 1970; Bernays, 1972*b*). Bernays assigns larval muscles to four useful categories: (1) permanent muscles, which are present throughout the life cycle, (2) first accessory muscles, which are present in hatchlings but atrophy during the first instar, (3) secondary accessory muscles, which are characteristic of nymphs but break down in adults, and (4) muscles which are rudimentary or absent at the beginning of postembryonic development and become functional only in adults. In the desert locust *Schistocerca*, a number of muscles fall into each of these categories. For instance, the pterothoracic dorsal longitudinal muscles (category 1) are employed in hatching, become essentially non-functional in nymphs, and develop into major indirect flight muscles of the adult (Bernays, 1972*b*; Thomas, 1954). In some hemipterans, they even disappear in early instars and redifferentiate later from an aggregation of myoblasts (Scudder & Hewson, 1971). Other dorso-ventral flight muscles, and genital muscles are very reduced in nymphs and develop late in the postembryonic period (Bernays, 1972*b*; Teutsch-Felber, 1970). In general, the anatomy of the musculature appears well correlated with behavioural development.

Anatomical descriptions have now been supplemented by ultrastructural, biochemical and physiological information. In the longitudinal flight muscles, differentiation is sharply accelerated at the time of imaginal moult in locusts (Brosemer, Vogell & Bucher,

1963; Vogell, 1965; Chari & Hajek, 1971; Walker, Hill & Bailey, 1970), and in crickets (Bocharova-Messner & Yanchuk, 1966; Chudakova & Bocharova-Messner, 1965). The amount of contractile protein increases markedly, while tracheation, sarcomere structure, and cellular organelles assume their mature conformation. Oxygen uptake and carbohydrate content are augmented, and the activity of key enzymes rises exponentially. These changes in muscle cells occur concurrently with development of co-ordination among flight neurons.

Few data have been obtained on excitable and contractile properties of muscle during development. In the cricket, *Acheta domesticus*, Chudakova & Bocharova-Messner (1965) have shown that by the 7th of 11 instars, flight muscles such as the dorsal longitudinal are capable of rapid, discrete contractions in response to stimulation of their nerve supply, and will respond with a tetanic contraction to stimulus trains. In *Schistocerca*, Tyrer (1969) recorded mechanical and electrical responses of abdominal dorsal intersegmental muscle to stimulation of motor nerves. A twitch could already be obtained in a 70% developed embryo, but the duration of contraction and relaxation shortened substantially by the time of hatching. Intracellular recordings from late embryonic muscle showed that this change was not due to alteration of the postsynaptic potential. This study is important both for its direct confirmation of anatomical evidence that nerve and muscle cells are functional in embryos, and also for demonstrating how early in development experiments with specific neuromuscular elements are feasible.

For the purposes of developmental neurophysiology, two important concepts can be derived from the studies of musculature. First, hatchling nymphs and late embryos have a very extensive complement of muscles, including at least the rudiments of almost every adult muscle. An advantage of this situation is that it should be possible to identify practically any motor neuron throughout the entire postembryonic period by its peripheral connections. Second, all through development there is a continually evolving musculature, with muscles maturing as they are needed and afterwards disappearing. In *Rhodnius*, this is carried to the extreme of differentiation and de-differentiation of the moulting muscles with each ecdysis (Wigglesworth, 1956; Warren & Porter, 1969). Therefore, the progression of nymphal behaviour reflects profound structural and physiolygical changes which require continual assembly and dissolution of neuromuscular elements.

Developmental neuroanatomy

The development of insect nervous systems has been reviewed recently by Edwards (1969) and Nuesch (1968). Some points concerning postembryonic development deserve emphasis. By hatching, mitoses can no longer be detected in neuroblasts of segmental ganglia in crickets (Panov, 1966; Gymer & Edwards, 1967) or locusts (Sbrenna, 1971). Cell body counts during development confirm that in these ganglia, all neurons are present at hatching. In the brain, and in segmental ganglia of mole crickets with delayed maturation (Panov, 1966), production of new neurons does continue during postembryonic development.

Although the full neuron cell body complement has been established in the embryo, segmental ganglia grow enormously after hatching. The terminal ganglion of *Acheta domesticus*, for example, increases in volume forty-fold by adulthood (Gymer & Edwards, 1967). While the number of neurons remains constant at about 100 motor neurons and 2000 interneurons, glial cells increase from 1000 to 17000. Despite the addition of glial cells, most of the volume increment is neuronal. Some of this is due to growth of the ganglion cortex, composed of cell bodies, but a larger proportion is contributed by the rapid expansion of the neuropile, or feltwork of neuron arborisations. The greatest relative growth is in the first four instars, where the portion of the ganglion devoted to neuropile increases from 42 % to 55 %. A similar growth pattern has been reported for the mesothoracic ganglion of *Schistocerca gregaria*, where the cortex volume increases 13-fold while the neuropile increases by 24-fold (Sbrenna, 1971) and for the prothoracic ganglion of the milkweed bug, *Oncopeltus fasciatus*, where the respective increases are 6-fold and 24-fold (Johansson, 1957). The disproportionate growth of cortex and neuropile suggests that a substantial part of a neuron's arborisation is not present at hatching and is constructed postembryonically. Since interactions between neurons are mediated by fibre to fibre contacts, many important events in the construction of neuronal circuits can be expected during this period.

In insects, the cell bodies of sensory neurons lie in the periphery and send axons into the central nervous system. Unlike motor neurons and interneurons, most sensory neurons develop postembryonically (Wigglesworth, 1953; Edwards & Palka, 1971; Svidersky, 1969). In hatchling *Schistocerca*, for example, no sensillae are present on the cuticle (Bernays, 1971). Sensory neurons differentiate from epidermal cells *de novo* throughout nymphal life. Consequently, important developmental events must also occur postembryonically in neuronal circuits whose activity is influenced by sensory input.

In summary, it appears that, like the musculature, the basic elements of the central nervous system are present at hatching, but that in many cases they have yet to undergo substantial differentiation.

Development of co-ordination among identified motor units

Behaviour requires a precisely timed sequence of contractions of the body musculature. Contractions are triggered by muscle action potentials which are initiated by arriving impulses from motor neurons, so that the behaviour reflects a co-ordinated firing of many motor neurons. This sequence of impulses is called a motor pattern or motor score. Studies have been undertaken recently to determine when and how such patterns arise in development.

The fundamental technique of these experiments has been electromyography. Insect muscles are innervated by a small, fixed number of neurons. For example, the subalar muscle, which depresses the wings in flight, is composed of two non-overlapping bundles of muscle fibres (Kutsch, 1969; Kutsch & Usherwood, 1970). Each of these is innervated by a single motor neuron. A fine, insulated wire pushed through the cuticle into one of the muscle bundles will record a muscle action potential whenever the unit contracts, and often the associated neuron impulse (Wilson, 1968; Fig. 1). Therefore, the exact firing pattern of identified motor neurons can be determined. Many wires can be placed in an animal without disrupting its activity, and a complete picture built up of the motor score underlying various behaviour patterns. After the fully co-ordinated pattern is known, recordings can be made from the same motor units at earlier stages to reveal how the pattern develops. Employing this procedure, several behaviour patterns have now been examined.

Cricket singing

Cricket songs are unusually stereotyped behaviour patterns which form part of a communication system. The repertoire of most species includes an aggressive song, mediating interactions between males, a calling song which attracts responsive females, and a courtship song facilitating copulation. The songs are composed of a sequence of sound pulses, each produced by closing the wings once. The development of the motor pattern underlying singing has been examined in the Australian field cricket, *Teleogryllus commodus* (Bentley & Hoy, 1970) and the European field cricket, *Gryllus campestris* (Weber, 1972).

Teleogryllus nymphs undergo nine, ten, or rarely eleven moults before reaching adulthood. When last instar nymphs are placed in situations which would elicit singing in adults, they do not produce movements of the wing buds in a song pattern. However, heat lesions in inhibitory areas of the brain (Huber, 1960) sometimes release fully developed calling, aggressive or courtship song patterns in nymphs. The calling song, for example, is a highly structured repeating phrase beginning with a chirp composed of 5–7 loud pulses separated by relatively long intervals. This switches abruptly into a trill of 6–10 more rapidly delivered soft pulses. One or two additional trills follow before the next chirp. The contraction pattern of wing muscles in the last instar nymph is identified as calling by (i) the number of bursts in the chirp and trill, (ii) the duration of interburst intervals, (iii) alternate firing of wing opener and closer muscles, and (iv) the presence and timing of the switch from long to short intervals (Fig. 1*a*). Consequently, last instar *Teleogryllus* nymphs appear to have completed neuronal networks generating song patterns, but premature activation of these circuits is prevented by inhibitory brain centres.

A similar situation has been reported for *Gryllus campestris* (Weber, 1972). Calling or courtship song could not be elicited from nymphs, and in forty cases of aggressive interaction, only two resulted in attempts to stridulate. Recordings from the wing opener and closer muscles showed that during these attempts, the motor pattern produced by last instar nymphs was similar to that of the adult aggressive song (Fig. 1*b*). Again, this indicates that the neuronal circuit generating the behaviour has been completed before the imaginal moult but is somehow suppressed.

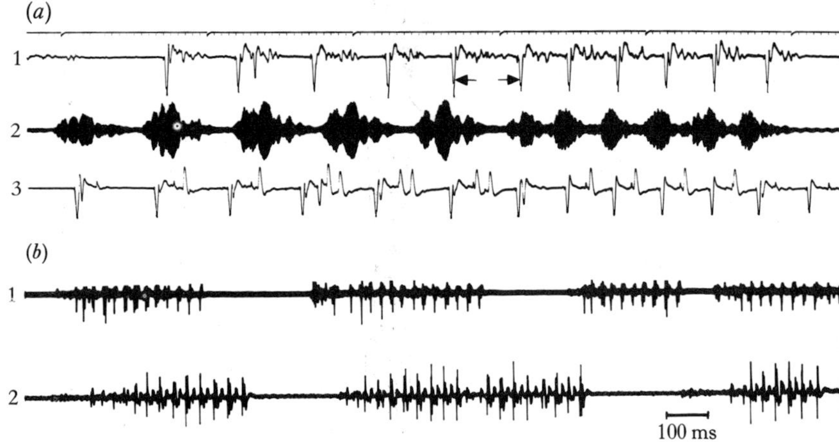

Fig. 1. Cricket singing: (*a*) Comparison of the motor output of a last instar *Teleogryllus commodus* nymph (1 & 3) with the calling song pulse pattern of an adult (2). (1) second basalar muscle (wing opener); note how closely the number and spacing of nymphal muscle potentials correspond to the adult pattern. (3) push–pull recording from the subalar muscle (downward spikes; wing opener) and remotor muscle (upward spikes; wing closer) demonstrating reciprocal firing of antagonists (closing movement produces sound pulse). Arrows indicate the switch (long intervals to left; short intervals to right) in both adult and nymphal patterns from long to short intervals between pulses. Time calibration: 5 ms per small division. (From Bentley & Hoy, 1970.) (*b*) Comparison of motor pattern underlying aggressive song of an adult *Gryllus campestris* (1) with the corresponding pattern of a last instar nymph (2). Action potential sequences from the mesothoracic subalar muscles are similar. (From Weber, pers. comm.)

Cricket flight

Orthopteran flight is produced by active beating of both the fore- and hindwings. The underlying motor pattern is characterised by: (1) rhythmical firing of flight motor neurons at the wingbeat frequency, often in multiple spike bursts, (2) approximately synchronous firing of synergistic neurons in the same segment, (3) approximately alternate firing of antagonistic units in the same segment, and (4) a lead of hindwing units over corresponding forewing units by about a third of a wingbeat cycle (Wilson, 1968). In addition to the pterothoracic musculature, many other parts of the body are involved in flight. An aerodynamic posture is assumed involving pointing the antennae straight ahead, drawing the fore-

and middle legs up close to the body, stiffening the abdomen for steering, and extending the hindlegs straight backward. *Teleogryllus commodus* flies well and conforms to this description.

In marked contrast to song patterns, elements of flight behaviour can be elicited from *Teleogryllus* nymphs early in postembryonic development (Bentley & Hoy, 1970). 7th instar nymphs, which must still undergo four moults before adulthood, will assume and hold the flight posture for short periods when suspended in a wind tunnel. As development continues, it becomes easier to initiate and maintain this posture.

Electrical recordings from the fore- and hindwing elevator and depressor muscles at each larval stage reveal a progressive development of the flight pattern. The first elements appear in the 7th instar, where short spike trains at about half the flight frequency can be recorded from hindwing depressor units. In the next stage, longer bouts of rhythmical activity occur with occasional two spike bursts. By the 9th instar, forewing motor neurons are also active at the proper frequency but often with incorrect relative timing. Hindwing depressor units fire in still longer trains, and the number of impulses in multiple spike bursts is equivalent to that seen in adults. In the final nymphal stage, all four sets of muscles are active and maintain the correct co-ordination, although the flight frequency remains low (Fig. 2). Therefore, the neuronal network underlying flight is also essentially complete before adulthood.

Gryllus campestris has reduced hindwings which cannot support flight. However, a pseudo-flight pattern which lacks the characteristic lead of hindwing over forewing units can still be generated (Kutsch, 1969). The development of this pattern is remarkably similar to that of *Teleogryllus* flight (Weber, 1972). In the fourth from last instar, both flight posture and the first elements of the motor pattern can be elicited. Recordings from wing elevator and depressor muscles show a few dozen rhythmical potentials at about two thirds of the flight frequency. The duration and frequency of flight increase progressively, with 8th instar nymphs flying up to a minute, 9th instar nymphs for two minutes, and last instar nymphs as long and rapidly as adults. The motor pattern of last instar nymphs features alternation of antagonists and synchronous firing of homologous mesothoracic and metathoracic units in the expected pseudo-flight co-ordination.

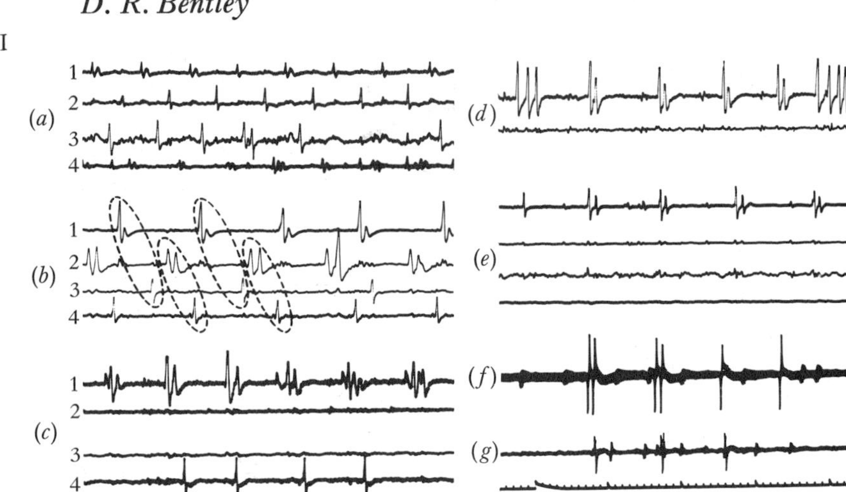

Fig. 2. Cricket flight: I. Progressive development of the flight pattern in *Teleogryllus commodus* nymphs. Recordings are muscle action potentials reflecting the discharge of single motor neurons. In each record (*a–g*), the top trace is the hindwing subalar muscle (wing depressor); the second trace (if any) is the hindwing remotor muscle (wing elevator); the third trace (if any) is the forewing subalar muscle; the fourth trace (if any) is the forewing remotor muscle. Recordings were made from all four muscles in each animal, but traces were deleted if the muscles were inactive in flight (accurate electrode placement could be confirmed because the muscles are bifunctional and used in walking). In the adult pattern, the wing depressors and elevators alternate in each segment, and the hindwings lead the forewings; broken lines (*b*) indicate this phase lag. (*a*) adult; (*b*) last instar; (*c*) second to last instar; (*d*) second to last instar; (*f*) third to last instar; (*g*) fourth to last instar; (*e*) this nymph, while also a second to last instar, will become an adult one moult earlier than the other animals of the figure, that is, after the ninth moult rather than the tenth. The figure illustrates the gradual emergence of the adult motor pattern during nymphal development; key features are (i) appearance of short bursts in the hindwing depressors, (ii) generation of sustained bursts, (iii) recruitment of antagonists and of the forewing motor units, (iv) development of the adult burst frequency. Note improper co-ordination of forewing elevator units in second to last instar (*c*). (From Bentley & Hoy, 1970. Copyright 1970 by the American Assoc. Advancement of Science.)

II. Flight pattern of DLM units in last instar *T. oceanicus*. (1) right metathoracic dorsal longitudinal muscle; (2) left metathoracic dorsal longitudinal muscle (wing depressor); (3) right promotor muscle (wing elevator). As indicated by

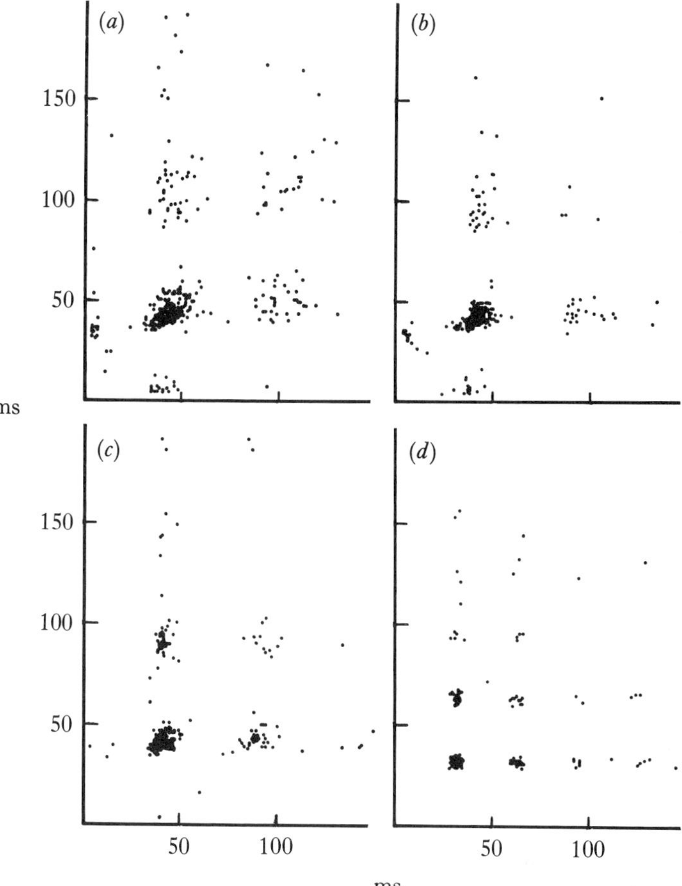

Fig. 3. Joint interval histogram demonstrating improvement of the flight motor pattern of *Gryllus campestris* nymphs with flight experience. The degree of clustering corresponds to the stability of the wing-stroke rhythm (method described in Roderieck, Kiang & Gerstein, 1962). Points represent the interspike interval of the mesothoracic subalar muscle (abscissa, n; ordinate, $n+1$). *a–c*: first, second, and third flights of a second to last instar nymph. *d*: first flight of an adult. (From Weber, 1972.)

---

synchronous firing of contralateral homologues and by alternation with antagonists, the DLM units are properly co-ordinated by the last instar. Following this recording, the DLM motor neurons were filled with cobalt dye and shown to conform to the adult structure. Time calibration: 5 ms per small division.

Weber (1972) has also examined the effect of use or experience on flight co-ordination. If 8th instar nymphs are placed in a windstream, units of the remotor muscle, a wing elevator, generate a burst of spikes followed by a period of rhythmical firing. In the first flight, the variance of the interspike interval, which corresponds to the interval between wingstrokes, is high compared to the adult. However, in the second flight of the same individual, and in additional subsequent flights the variance decreases considerably so that it becomes much less than that of inexperienced animals at the same stage of development (Fig. 3). Evidently, use of this neuronal circuit produces some rapid changes in network features responsible for the stability of the rhythm.

To review, elements of the cricket flight pattern appear as early as the 7th instar and build progressively throughout the last third of nymphal development. Before the imaginal moult, co-ordination of the motor units involved in both flight and song patterns is essentially complete.

Locust flight

The ontogeny of flight has been examined in three species of locusts, and follows a similar, but slightly delayed, course. Locusts have only five instars, about half as many as crickets, and generally are much better fliers. In *Schistocerca*, the first evidence of flight posture occurs in first instar nymphs, which will hold their legs spread for a few moments if placed in a wind stream (Kutsch, 1971). Second to fifth instar nymphs will all assume the true flight posture, but older nymphs take it more readily and hold it longer. Flight motor units are activated as early as the fourth instar, about the same relative time in postembryonic development as the corresponding units become active in crickets. Passages of firing are brief but often quite regular with a mean of about 20 Hz. In the fifth instar, more units are active with complex and variable firing patterns. The most typical situation is a steady discharge at approximately 20 Hz, with some double firings. However, proper co-ordination for flight has not been established at this stage. Synergistic units are not strongly coupled and antagonists do not fire alternately (Fig. 4*a*).

On the day of the imaginal moult, young adults will already flap their wings at 10 Hz or roughly half the adult frequency (Kutsch, 1971). Recordings of wing elevator and depressor units show

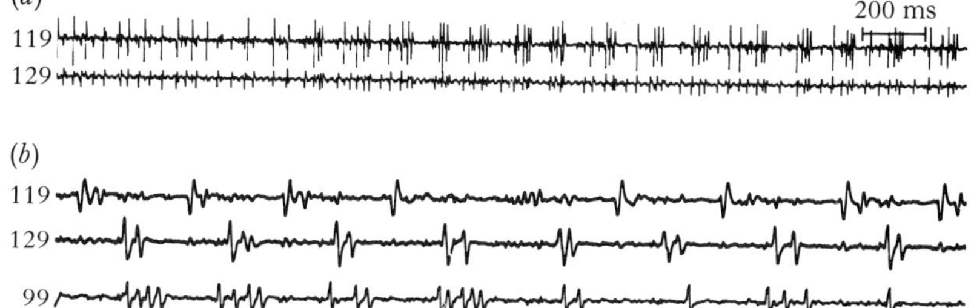

Fig. 4. Locust flight: (*a*) Activity of the metathoracic remotor muscle (119, wing elevator) and the metathoracic subalar muscle (129, wing depressor) of a fifth instar *Schistocerca gregaria* nymph during suspension in a windstream. Patterned activity (see text) is more apparent in the remotor muscle. (*b*) Flight pattern of adult *Schistocerca* on the day of the final moult. The antagonistic remotor and subalar muscles fire alternately; however, the phase lead of the metathoracic subalar (129) over its mesothoracic homologue (99) has not been established. (From Kutsch, 1971.)

rhythmical firing with the frequent two-spike bursts characteristic of adults. Synergists fire quite synchronously, and antagonists maintain alternation even at different flight frequencies. Hindwing motor units usually lead forewing units, but not as much as in the fully mature pattern. Therefore, a characteristic although somewhat imperfect flight pattern has been established (Fig. 4*b*).

During the first few weeks of adult life, a number of steady improvements occur which result in a fully co-ordinated pattern. In newly moulted animals, occasional lapses of relative timing, such as a loss of antagonist alternation, are seen and these faults are gradually eliminated. The variance in interburst interval, which is quite high, is reduced until the wing stroke rate becomes very regular. In the two-spike bursts, the interval between impulses is about twice as long as it should be, and is decreased from 20 ms to 10 ms. The lead of hindwings over forewings is stabilised and increased until it reaches 30 degrees. Finally, the frequency of wing beating accelerates from 10 Hz to the 20 Hz rhythm found in mature adults (Kutsch, 1971). Michel (1971) also reports a progressive increase in the readiness of animals to fly and in the duration of flights. A similar relationship between age and flight duration is found in other insects, including cockroaches (Farnworth, 1972), *Oncopeltus* (Dingle, 1965), and army ants (Schneirla,

1948). Therefore, a prolonged and substantial period of development of this behaviour pattern occurs after the imaginal moult.

Wingbeat frequency

From these several features of the flight pattern which undergo marked development in the adult, Kutsch (1971, and personal communication) has examined the increase in wingbeat frequency in more detail. Recordings from *Locusta* indicate that the increase takes an exponential form (Fig. 5). Exponential increases have also been recorded in *Drosophila* (Chadwick, 1953), *Phormia* (Levenbrook & Williams, 1956), mosquitoes (Roth, 1948), and cockroaches (Farnworth, 1972). In *Locusta migratoria*, daily frequency measurements were made during the first month of adult life (Kutsch, personal communication). Development of the flight motor pattern in *Locusta* is similar to *Schistocerca*, but the peak frequency is higher (24 Hz). The measurements fit the exponential function: $f_t = F-(F-f_0)\mathrm{e}^{-kt}$, where f_t is the daily frequency, F is the terminal frequency, f_0 is the initial frequency, t is the days after imaginal moult and k is the rate constant. For males, $f_t = 23.8\ (23.8-9.8)\mathrm{e}^{-0\cdot12t}$. This calculation indicates that the theoretical starting frequency of the flight oscillator is slightly less than 10 Hz, and that the adult developmental period is quite long, so that even after 30 days an increase of 0.5 Hz is still to be expected.

The effect of temperature on the rate and end points of wingbeat frequency maturation was investigated by raising *Locusta* from the fifth instar onwards in cold (23 °C), intermediate (31 °C) and hot (37° C) cages. Frequency measurements at a standard temperature showed differences in slope, with animals from warmer cages maturing faster ($Q_{10} = 2.7$), but all three curves had about the same initial and terminal frequencies. Kutsch (personal communication) suggests that although the time span of development can be influenced by external temperature, the absolute frequency of flight rhythm is pre-programmed by other factors, possibly genetic.

One factor which might be critical in the frequency increase is flight experience as opposed to maturation time. Kutsch (1971) has tested this with two groups of newly moulted *Schistocerca*: individuals in the control or naive group were flown only once, but at different times after reaching adulthood. The test group was given 5–10 minutes of flight experience daily. Comparison of the

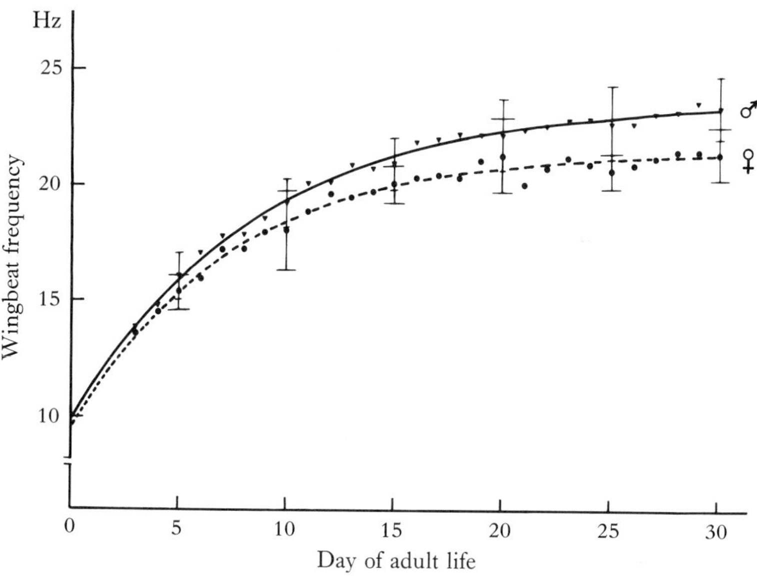

Fig. 5. The exponential increase in wingbeat frequency of *Locusta migratoria* during the first few weeks of adult life. Points are the average of daily measurements from 9 males and 6 females. (From Kutsch, personal communication.)

flight frequency of animals from the naive and the experienced groups on each day after ecdysis showed no significant differences. Therefore, experience does not appear to be important in the increase in flight frequency.

In *Locusta*, flight frequency changes have been followed through the entire adult life span. For most animals, three different periods can be distinguished: (1) the initial period of exponential frequency increase, (2) a plateau of frequency stability, with variations of ± 2 Hz, and (3) a terminal phase of increased frequency variation, up to 4 Hz, superimposed on a slow decline in mean frequency. On the day of death, some animals show an abrupt frequency decline but in others the performance is still quite regular (Kutsch, personal communication). The changes which this parameter of the flight pattern exhibits emphasise that development does not cease with the imaginal moult but continues well into adulthood.

The transition to adult co-ordination

One of the most interesting phases in development of the flight pattern in *Schistocerca* is the transition from unco-ordinated firing of motor neurons in the fifth instar to co-ordinated firing in the young adult. Altman (personal communication) has examined this transition in more detail in the Australian plague locust, *Chortoicetes terminiferous*. This locust has five or six instars and development of the flight pattern closely parallels that of *Locusta* and *Schistocerca*. From the first instar, nymphs respond to a wind stimulus and the correct flight posture is usually assumed by the fourth or fifth instar. Rhythmical activity of flight motor units is already present in the fourth and fifth instars, but all muscles in both segments fire approximately synchronously. There is no abrupt change to the adult pattern of antagonist alternation. By the middle of the fifth instar, some individuals produce brief passages of alternation, while in others this feature of co-ordination is not established until after the final moult. By three to six hours after the moult, most animals show alternation, but firing is irregular and the wings do not make a full excursion. Phase relationships improve gradually until a stable pattern is achieved by the end of 24 hours. There is some evidence that experience plays a role in this process. Wingbeat frequency is low at first (14–16 Hz), undergoes a rapid increase to 20–22 Hz after 3–5 days, and then increases slowly to 25–33 Hz during the first three weeks of adult life. Therefore, even in the critical period for development of co-ordination, the changes seem to be steady and continuous rather than step-like.

Factors underlying the development of co-ordination

The data obtained by the electromyogram technique are the occurrence and relative timing of discharge of identified motor neurons. These discharge patterns have been shown to undergo marked change during postembryonic development, but what do the changes mean in terms of properties of the nervous system? One possibility is that they do not reflect the actual construction of a functional neuronal network, but are due to a gradual withdrawal of inhibition of previously constructed circuits. The neurons generating cricket calling-song do appear to be under suppression in the last instar, for removal of inhibitory influences exposes a fully co-ordinated discharge pattern. For the flight circuit, this possibility seems quite remote. In both crickets and locust, the flight pattern develops over a long period of postembryonic life. Simple elements appear first and flight co-ordination, duration, frequency, and posture are gradually and continuously improved. There is no evidence for declining inhibition, and the most attractive hypothesis is that the change in motor unit behaviour directly reflects maturation of elements of the flight neuronal network. These elements include sensory, motor, and interneurons, and for each cell, differentiation of internal properties and external connectivity. Some information on the respective roles of these factors is available.

The wind-sensitive hairs or trichoid sensillae on the head are important in orientation, initiation, maintainance and frequency of locust flight (Wilson, 1968; Camhi, 1969). The changes in the last three factors seen in development of both cricket and locust flight could be caused by maturation of these sensillae. In *Locusta migratoria*, Svidersky (1969) has reported an ontogenetic change in both the number and physiology of these receptors. A few hairs are present in the first instar but many additional cells differentiate throughout the larval period. In the young cells of first and second instars, the response to a wind stimulus is a brief train of impulses. This phasic response is extended in the third and fourth instars until by the last instar many units respond in the adult, tonic fashion. The development of this sensory system must make an important contribution to the maturation of the flight response.

The four stretch-receptor neurons, one coupled to each wing, are critical in maintaining flight frequency in adult locusts (Wilson, 1968). Destruction of these sensory cells reduces frequency to about half the normal level. The absence of this input in nymphal *Teleogryllus* may be an important factor in the low frequency of the flight pattern (Bentley & Hoy, 1970), although in adult *Gryllus campestris*, a non-flying species, these receptors have lost their influence on 'flight' pattern rhythm (Möss, 1971). In last instar nymphs of *Schistocerca*, Kutsch (1971) has shown that all four receptors fire normally if the wing bud is pulled, that they are active in the first adult flights, and that cutting them at any time during a flight results in the expected drop in frequency (Fig. 6). Therefore, this reflex is operative immediately after the imaginal moult, and cannot account for the frequency increase observed during the first weeks of adulthood.

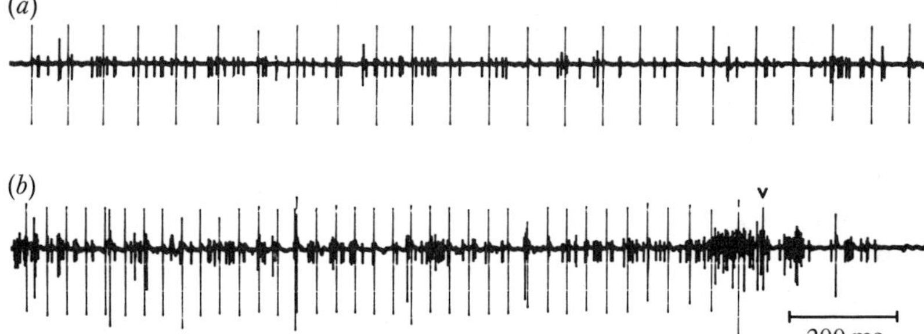

Fig. 6. Activity of sensory neurons mediating flight reflexes in a fifth instar *Schistocerca* nymph. Large spikes are from the stretch receptor neuron and small spikes are from the chordotonal organ. (*a*) resting activity (*b*) response to pulling the wing bud (stretch is released at v). The figure shows that sensory elements of the flight network are also physiologically active before adulthood. (From Kutsch, 1971.)

The development of patterning in the output of a neuron can be resolved into two elements: the first includes characteristics like the average firing frequency, the amount of multiple spike bursting, and the onset of firing in ontogeny. These are governed both by internal properties like threshold, refractory period, and responsiveness, and also by input. Experiments separating internal from external effects on single cells would be very useful. There is some evidence that with respect to these properties, the sequence of development of different types of motor neurons is determinate, since in both crickets (Bentley & Hoy, 1970) and locusts (Altman, in preparation) depressor units appear to become active before elevators. The other element is the relative timing of impulses in different units. In crickets and locusts, correct timing emerges secondarily from a period of unco-ordinated firing. Since co-ordination is produced by interactions between neurons, this development should correspond to the construction of functional synapses. Altman (in preparation) suggests, for example, that the gradual achievement of alternation between antagonists is caused by new inhibitory synapses. There is no indication so far that the order in which synapses are established is pre-determined. In view of the relationship between physiological maturation and structure, it should be informative to investigate the anatomy of these neurons during development.

Structural development of identified neurons

I have recently examined the structure of some motor neurons and sensory neurons involved in flight and stridulation of *Teleogryllus oceanicus* at most stages of postembryonic development. The configuration of individual cells within the thoracic ganglia was made visible with cobalt dye (Pitman, Tweedle & Cohen, 1972) introduced into the neurons by axonal iontophoresis (Iles & Mulloney, 1971). This method has several useful features: the cells are first photographed in stereo-pairs of cleared whole mounts to preserve three-dimensional information. Subsequently, optical sections and also ultra-thin sections are taken from the same preparation so that detailed information on anatomy and ultrastructure is obtained. This procedure reveals an exceedingly complex arborisation even in motor neurons that must be among the simplest in the ganglion (Plate 1). Analysing this structure in terms of connections to other cells requires simultaneous filling of several identified neurons. Such multiple fills are quite a challenge with microelectrodes, but relatively simple through axonal iontophoresis, especially since associated motor neurons and sensory neurons often have axons in the same peripheral trunk. Since the cells are filled through the peripheral nerve rather than by microelectrode penetration, it is possible to use smaller ganglia and, therefore, earlier developmental stages. In the cricket, identified neurons have already been filled down to the third instar, and it should be possible to fill embryonic motor neurons which are connected to their muscles. Finally, the unusual reliability of the technique makes it feasible to record routinely the physiological performance of a cell and then to determine its structural development.

Using this method, representative elements of the motor and sensory systems mediating wing movements were examined. Motor neurons innervating the pterothoracic dorsal longitudinal muscles (DLM) were selected because they are the only major flight muscles in crickets which are not also connected to the legs. Therefore, the arborisation is uncomplicated by structures involved in walking and, since the muscles appear to be used little during nymphal life, might develop relatively late. The sensory neurons investigated were the four stretch receptors, and cells innervating wing cuticular sensillae. This combination of elements has the advantage for axonal iontophoresis that all have axons in the wing nerve (NICD; terminology after Campbell, 1961). This trunk also sends branches to the dorsal oblique muscle and apparently to dorsal cuticular receptors (Fig. 7). These branches were dye-filled in various unilateral, bilateral, and multisegmental combinations in 51 male crickets, including 17 adults and the remainder in each nymphal stage down to third instar.

Filling the nerve revealed seven major classes of neurons (Fig. 7, Plate 1), five with central somata, presumably motor, and two without central somata, presumably sensory (additional cells sometimes filled which are not described here). The dorsal longitudinal muscles of both segments are each innervated by five motor neurons, as previously demonstrated in *Schistocerca* (Neville, 1963; Guthrie, 1964; Bentley, 1970). Four of these neurons (DLM class 1) originate in the ganglion of the thoracic segment anterior to that containing the muscle (Fig. 7). These cells have ipsilateral somata and ipsilateral arborisations. The fifth motor neuron (DLM class 2) originates in the ganglion of the same segment. Its arborisation is primarily ipsilateral to the innervated muscle, but the soma is contralateral. In *Teleogryllus* this soma is somewhat lateral to the position shown by Guthrie (1964) and in the ganglion map by Bentley (1970) for the corresponding *Schistocerca* neuron; Tyrer & Altman (in preparation) now report that the locust cell is also in the lateral position. Class 3 is a group of about four, laterally placed, small somata which innervate the lateral oblique muscles. If these larval muscles atrophy in the adult, as in *Locusta* (Wiesend, 1957) and probably *Schistocerca* (Bernays, 1972*b*), it is interesting that the somata, axons, and some arborisation are still present after the imaginal moult. Class 4 includes a single, large, dorso-medial cell body with a centrally bifurcating axon that sends branches out of both wing nerves. This remarkable neuron undergoes a striking change at the imaginal moult. Class 5 consists of a pair of tiny, ventral, midline somata in each of the three thoracic ganglia. The function of these neurons is unknown. Of the cells without central somata, class 6 is represented by an axon arriving from each wing nerve branch $1D_2$. This fibre bifurcates in the periphery, sending one axon into the ganglion of the same segment and another to the next anterior ganglion. The latter branch has not filled well, but the branch in the same segment weaves along the entire length of the ganglion. Presumably, these are the axons of the four stretch receptors (Möss, 1971). The final class, 7, is a large group of small

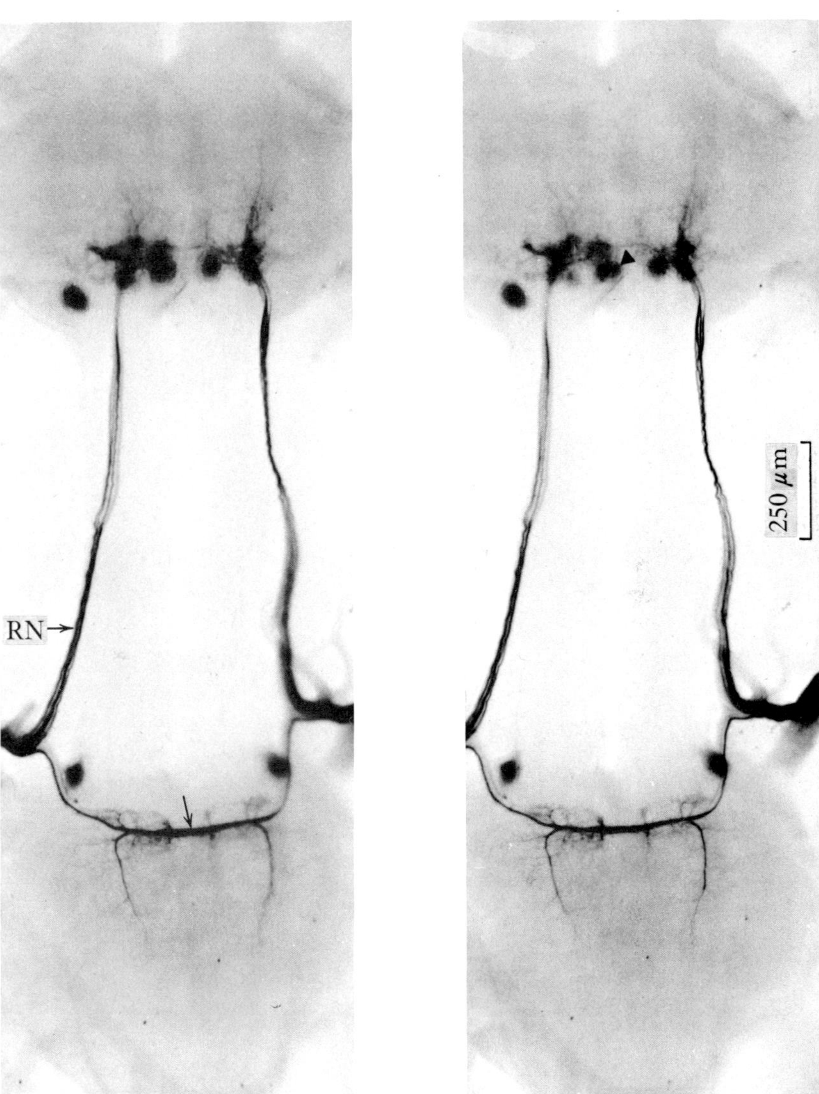

Plate 1. The 10 motor neurons innervating the metathoracic dorsal longitudinal muscle of an adult cricket, *T. oceanicus*. Cells were filled with cobalt dye through axons in the wing nerve, and photographed in stereo pairs from a cleared whole mount (ganglia tilted to right). Arborisations and somata of the pair of class 2 DLM neurons are in the metathoratic ganglion (bottom). Note the proximity of the decussating axons (arrow). Ipsilateral arborisations and somata of the 8 class 1 DLM neurons are in the mesothoracic ganglion (top); the triangle indicates the midline branch. RN, recurrent nerve.

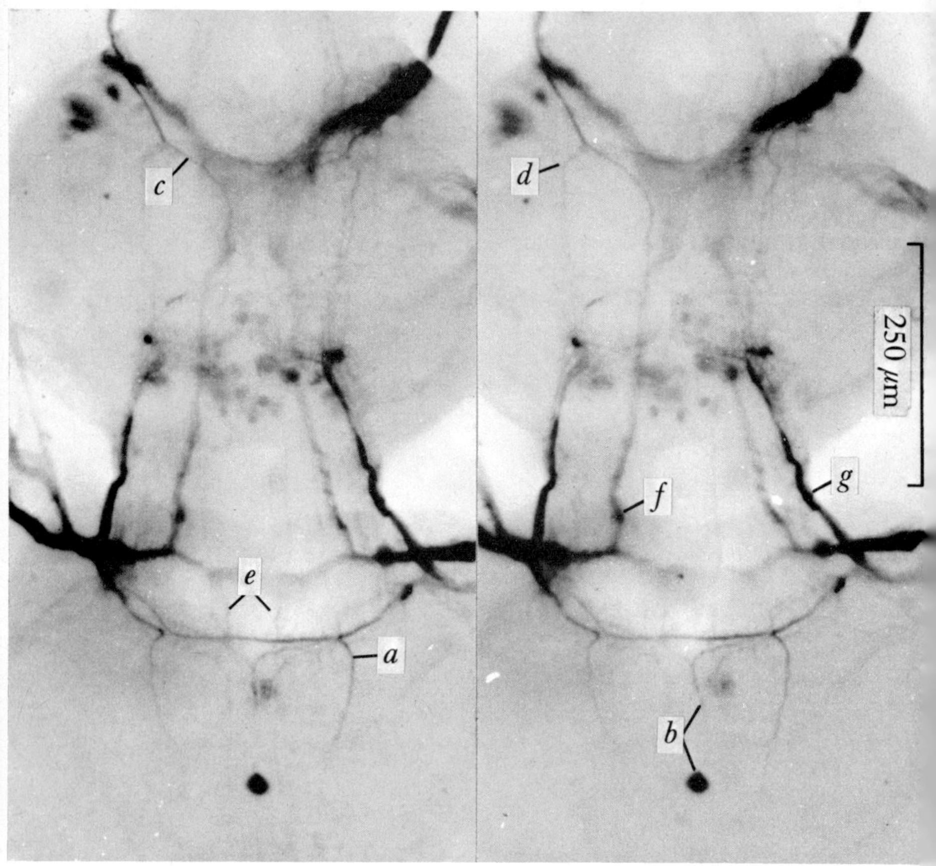

Plate 2. Stereo photograph of cobalt-filled wing-nerve sensory and motor neurons in the mesothoracic (top) and metathoracic (bottom) ganglia of a sixth instar male *T. oceanicus* nymph: (*a*) class 2 DLM neuron innervating muscle 112; fewer secondary branches appear than in the adult (Plate 1), (*b*) dorso-medial neuron, (*c*) mesothoracic stretch receptor axon, (*d*) branch of class 2 DLM neuron from muscle 81 which enters the arborisation of class 1 DLM neurons from muscle 112, (*e*) possible interganglionic DLM fibre, (*f*) axons of wing cuticular receptors (see text), (*g*) class 1 DLM neurons from 112. Note the relatively short length of the interganglionic connective compared to the adult (Plate 1).

axons from the wing sensory nerve (NIC). After entering through root 1, this bundle plunges to the bottom of the ganglion and splits into two tracts. The first crosses to the midline where it meets contralateral homologues; the second runs through the ipsilateral connective to contact intersegmental homologues of the adjacent anterior ganglion. Although receptors in this nerve do influence wing movements (Wilson, 1968; Möss, 1971), they were not seen to interact directly with stretch receptor or motor neurons.

Structural concomitants of co-ordination

When appropriate sets of neurons are filled with dye, structural concomitants of some physiological interactions can be seen. For example, flight co-ordination involves (i) synchrony of synergists, (ii) alternation of antagonists, (iii) an intersegmental phase shift, and (iv) sensory reflexes (singing co-ordination is similar except that only one segment is involved). Synchrony of synergists among the DLM neurons requires positive coupling between (1) contralateral homologues, (2) motor pool-mates in the same ganglion (DLM class 1) and (3) motor pool-mates in different ganglia (DLM class 1 with DLM class 2). The contralateral DLM class 2 neurons have ample opportunity to interact since their decussating axons run together for at least a third of the ganglion width. In fact, it is hard to imagine that interaction could be avoided. This juxtaposition is not simply fortuitous since the non-decussating DLM class 1 neurons send a major branch to the midline to meet their contralateral homologues. This branch is also found in *Schistocerca* neurons (Bentley, 1970). There is no direct, central connection between DLM neurons from different ganglia which innervate the same muscle; however, each motor neuron has a branch in the same through tract so the coupling is probably mediated by an interganglionic interneuron. Arborisations of motor pool-mates in the same ganglion lie in close register through several degrees of branching and offer numerous sites for interaction (Plate 1; Bentley, 1970). Therefore, several cases of synergistic coupling do have structural counterparts (Fig. 7; Plate 1).

Although the primary antagonists of DLM neurons have not been dye filled, antagonistic flight motor neurons in *Schistocerca* do have interdigitating branches (Bentley, 1970). In crickets, the mesothoracic DLM is actually split into two antagonistic bundles (Bentley & Kutsch, 1966; Kutsch, 1969), and of course the

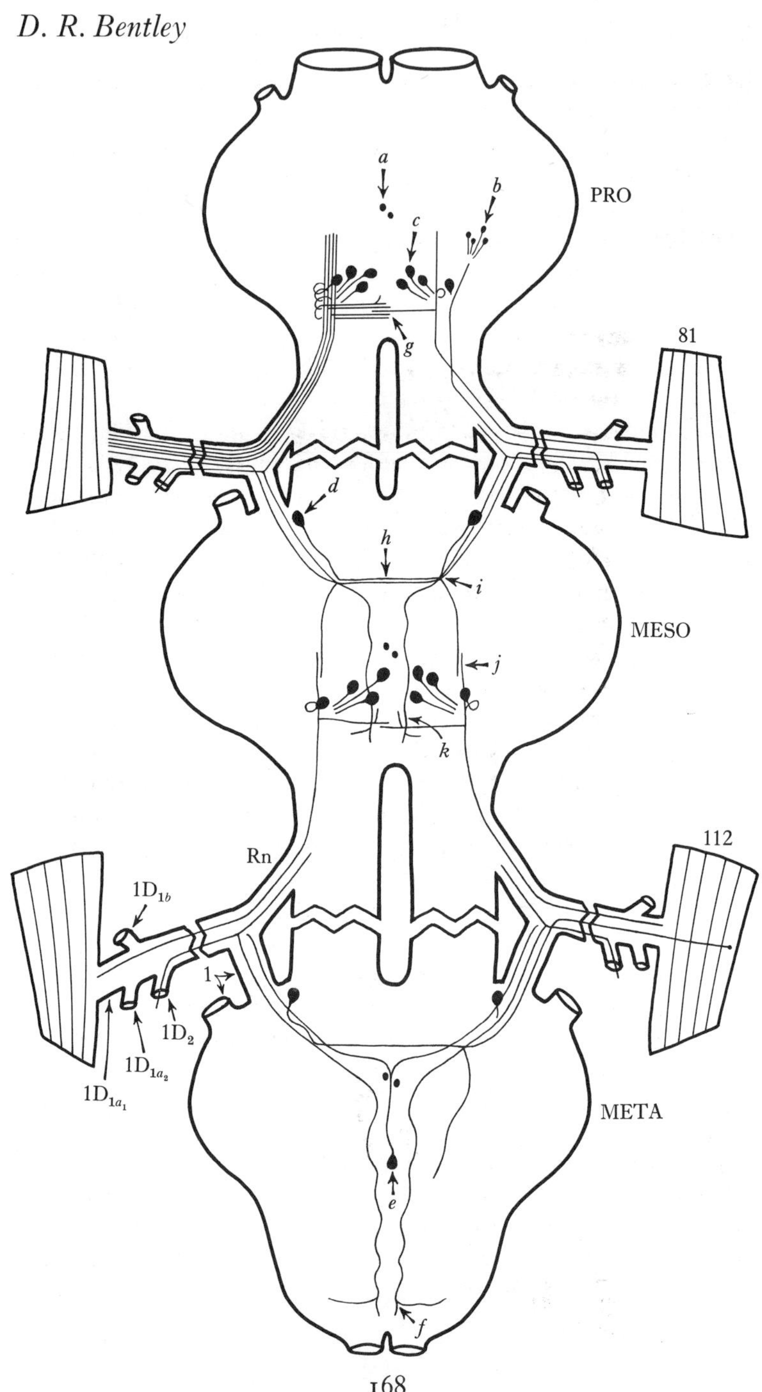
a
b
PRO
c
g
81
d
h
i
MESO
j
k
112
Rn
1D1b
1
1D2
1D1a2
1D1a1
META
e
f

neurons innervating these have numerous sites for possible interaction.

The intersegmental phase shift requires co-ordination between metathoracic DLM motor neurons and mesothoracic DLM neurons. Both motor pools have units in the mesothoracic ganglion, where a major anteriorly running branch of the neurons innervating the metathoracic muscle interdigitates with a large, posterior branch of the neurons innervating the mesothoracic muscle.

Finally, an influential sensory input to the flight motor system is provided by the stretch receptors. In the neuropile, stretch receptor axons first pass close to the DLM units of the same segment, and then continue on into the arborisation of DLM units innervating the adjacent posterior segment. Whether the proximity of these flight neuron branches represents no interaction, mutual access to an interneuron, or direct synaptic contact has not been determined, although the appropriateness of the structure is certainly suggestive. Electron micrographs support the optical

---

Fig. 7. Diagram of primary branches of the wing nerve sensory and motor neurons involved in flight and stridulation, showing development by the sixth instar in *Teleogryllus oceanicus*. Six cell classes are shown (*a–f*): (*a*) paired, midline, ventral somata, found in each thoracic ganglion, (*b*) somata of motor neurons innervating the dorsal oblique muscle (82); only one axon and one somata cluster are shown, but homologous cells are present bilaterally in each ganglion, (*c*) class 1 DLM motor neurons innervating the mesothoracic muscle; both cell bodies and arborisations are ipsilateral to the muscle. Homologous neurons are present bilaterally in the pro- and mesothoracic ganglia, (*d*) class 2 DLM motor neurons innervating the mesothoracic muscle; the soma is contralateral to the muscle. Homologous neurons are present bilaterally in the meso- and metathoracic ganglia, but in the latter ganglion only branches of the left cell are shown, (*e*) metathoracic dorsomedial neuron, with an axon in both wing nerves, (*f*) arborisation of stretch receptor neuron; homologous cells occur bilaterally in both pterothoracic ganglia. Sites of close proximity between fibres which could mediate observed physiological interactions are indicated (*g–k*): (*g*) interdigitation of contralaterally homologous DLM class 1 neurons (shown only for selected cells). Note also that arborisations of ipsilateral homologues are in register (shown only for left prothoracic units), (*h*) decussation of DLM class 2 neurons; fibres run together for at least a third of the ganglion width (shown only for mesothoracic units), (*i*) common intersection of both DLM class 2 neurons and the stretch receptor (also present in metathorax), (*j*) overlap of branches of DLM neurons innervating muscles in different segments, (*k*) crossover of stretch receptor with DLM neurons from next posterior segment. In general, the structural framework of these neurons appears well established by half-way through nymphal development.

evidence but have not yet shown synaptic structure (Richard Clark, this laboratory). Synaptic contact is a possibility since both excitatory (Wilson, 1968; Bentley, 1969) and inhibitory (Levine, 1972) direct interactions between insect flight motor neurons occur. Whatever the exact role of neuron structure, it must be completed before the onset of co-ordinated physiological activity.

Postembryonic structural development

In *Acheta domesticus*, the dorsal longitudinal muscles of both segments are present embryonically as five thin strands, one for each motor unit (Voss, 1911, 1912). In freshly hatched *Teleogryllus* nymphs, the wing nerve can be dissected to the muscle, and also appears in optical sections. Therefore, the DLM motor neuron somata and peripheral axons almost certainly arise embryonically. The earliest dye-fills of the wing nerve have been obtained in the third instar, where the right metathoracic DLM somata in the mesothoracic ganglion can be identified. By the fourth instar, the somata and major neuropile branches of all ten metathoracic DLM neurons filled. In the fifth instar, the mesothoracic DLM somata and primary branches are present as well as the metathoracic dorsomedial cell, the primary branches of the stretch receptors, and the wing sensory neurons of both pterothoracic ganglia. It should be noted that while successful dye filling demonstrates the presence of a structure, failure to fill does not necessarily prove its absence; consequently, any of these neural elements could be present even earlier in development.

By the sixth instar, half-way through nymphal development, all seven cell classes described for the complete system are present (Fig. 7, Plate 2), including the DLM units in all three ganglia, the stretch receptor and wing sensory neurons, and both groups of midline somata. Arborisations are well developed with at least the primary and secondary branches of DLM units and stretch receptors present. Tertiary and finer branches are not seen to be as full as in the adult, but whether this reflects incomplete maturation or incomplete filling is uncertain. All of the structural concomitants of co-ordination described above are present. Therefore, at least one instar before physiological activation of the flight neuronal network, the structural framework of appropriate motor and sensory units have been built (analysis of physiological activity and then structural development of DLM neurons of the same animal has been begun, and so far conforms to the pattern reported for other flight motor neurons; units do not show flight-related activity in the sixth instar and are fully co-ordinated by the last nymphal instar; (Fig. 2*b*).

At the imaginal moult, one marked structural change occurs in the circuit; the metathoracic dorsomedial neuron 'disappears' (as seen by this technique). It fills almost invariably in nymphs (Plate 2) but has not been observed in any of the 17 adult preparations. Since its axon is found in a nerve innervating sensory and motor structures used only in adult behaviour, an attractive hypothesis is that the dorso-medial neuron is involved in suppression of these behaviour patterns during nymphal life and is lost when no longer needed. Most characteristics of the cell correspond to those of a group of eight large, dorso-medial neurons recently described in adult locusts and cockroaches (Crossman, Kerkut, Pitman & Walker, 1971), which are notable as the only known insect neurons with electrically excitable somata. The dorso-medial cells of adults bifurcate centrally and then branch repeatedly, sending axons out of every motor trunk *except* the wing nerve (Crossman, Kerkut & Walker, 1971). Since the nerves receiving axons innervate only leg muscles, it is supposed that these neurons modulate posture or locomotion by influencing either motor or sensory structures. Peripheral inhibition, for example, is well known in leg musculature, but does not play a role in adult wing movements. Therefore, it seems possible either that the nymphal dorso-medial neuron is one of the group of eight, and simply loses the unneeded wing nerve axon while retaining others, or that there are nine dorso-medial cells in nymphs and the wing nerve cell is lost at the imaginal moult.

Summary and conclusions

Analysis of the maturation of identified neuronal and muscular elements generating adult behaviour of non-metamorphosing insects is a relatively recent undertaking. Information available so far is consistent with the following developmental scheme:

(1) Most muscles differentiate embryonically, although some remain rudimentary until late nymphal development.

(2) Cell bodies of neurons in segmental ganglia arise embryonically, and usually motor neurons innervate their muscles before hatching.

(3) Arborisations of neurons within the ganglia grow substantially during the first third of nymphal development (instars 1–4 in crickets; 1–2 in locusts). In the case of neurons employed in more than one behaviour, it is not known whether the arborisation grows as a whole or if branches grow first which are necessary for behaviour patterns appearing earlier in development.

(4) The basic framework of sensory and motor neurons mediating cricket flight is completed half-way through nymphal development (6th instar), one stage before activation of the flight circuit begins. Structural concomitants of a number of physiological interactions are present.

(5) Finer branches, dendritic spines, and synaptic ultrastructure have not been examined but presumably are laid down next, in the interval between the appearance of the cell framework and the onset of physiological activity.

(6) In the last third of nymphal life (from 7th instar in crickets; 4th in locusts), physiological activation or spiking of motor neurons, sensory neurons and presumably interneurons in the flight circuit can be elicited. This may reflect maturation of internal properties controlling excitability, or the development of input synapses.

(7) After a period of unco-ordinated firing, correct relative timing among flight motor neurons gradually emerges; this probably reflects the establishment of functional synapses between neurons. Flight experience accelerates the development of some features of co-ordination, but not others. Activation of these synapses might not have a detectable structural correlate, and might instead be triggered biochemically. Such changes could be revealed by a biochemical analysis of the appropriate cell bodies.

(8) The major features of flight and singing co-ordination are established in crickets by the last instar, one stage before the behaviour normally appears. In locusts, the corresponding phase is slightly later, usually just after the imaginal moult. However, adjustments in flight co-ordination, initiation, duration, and frequency continue for some time, and the fully mature pattern is not developed until well into the adult period.

In addition to the exclusively adult behaviour, insects show an extensive range of nymphal behaviour patterns which appear for varying lengths of time at different stages of development. The assembly and de-differentiation of neuromuscular machinery underlying these behaviour patterns offers many excellent opportunities for developmental study, and rapid advances in this field can be expected.

I thank Mrs Alma Raymond for technical assistance, and Dr Ronald R. Hoy for collaboration on the physiological studies of crickets. Support was provided by USPHS grants NS09074 and FR7006.

References

Bentley, D. R. (1969). Intracellular activity in cricket neurons during the generation of behavior patterns. *Journal of Insect Physiology*, **15**, 677–99.

(1970). A topological map of the locust flight system motor neurons. *Journal of Insect Physiology*, **16**, 905–18.

(1971). Genetic control of an insect neuronal network. *Science*, **174**, 1139–41.

Bentley, D. R. & Hoy, R. R. (1970). Postembryonic development of adult motor patterns in crickets: a neural analysis. *Science*, **170**, 1409–11.

(1972). Genetic control of the neuronal network generating cricket song patterns. *Animal Behaviour*, **20**, 478–92.

Bentley, D. R. & Kutsch, W. (1966). The neuromuscular mechanism of stridulation in crickets (Orthoptera: Gryllidae). *Journal of Experimental Biology*, **45**, 151–64.

Benzer, S. (1971). From the gene to behavior. *Journal of the American Medical Association*, **218**, 1015–22.

Bernays, E. A. (1971). The vermiform larva of *Schistocerca gregaria* (Forskal) – form and activity (Insecta, Orthoptera). *Zeitschrift für Morphologie der Tiere*, **70**, 183–200.

(1972*a*). The intermediate moult (first ecdysis) of *Schistocerca gregaria* (Forskal) (Insecta, Orthoptera). *Zeitschrift für Morphologie der Tiere*, **71**, 160–79.

(1972*b*). The muscles of newly hatched *Schistocerca gregaria* larvae and their possible functions in hatching, digging and ecdysial movements (Insecta: Acrididae). *Journal of Zoology*, **166**, 141–58.

Bocharova-Messner, O. M. & Yanchuk, K. (1966). Ontogenetic changes in the ultrastructure of the wing muscles of *Acheta domesticus* L. *Doklady Akademii Nauk SSSR*, **170**, 948–51.

Brookes, H. M. (1952). The morphological development of the embryo of *Gryllus commodus* Walker (Orthoptera: Gryllidae). *Transactions of the Royal Society of South Australia*, **75**, 150–61.

Brosemer, R. W., Vogell, W., & Bucher, T. (1963). Morphologische und enzymatische Muster bei der Entwicklung indirekter Flugmuskeln von *Locusta migratoria*. *Biochemische Zeitschrift*, **338**, 854–910.

Camhi, J. M. (1969). Locust wind receptors. I. Transducer mechanics and sensory response. *Journal of Experimental Biology*, **50**, 335–48.

Campbell, J. I. (1961). The anatomy of the nervous system of the mesothorax of *Locusta migratoria migratorioides* R and F. *Proceedings of the Zoological Society of London*, **137**, 403–32.

Chadwick, L. E. (1953). The motion of the wings. In *Insect Physiology*, ed. K. D. Roeder, pp. 577–614. New York: Wiley.

Chari, N. & Hajek, I. (1971). Metabolic differentiation of flight and jumping leg muscles of locust. *Physiologia Bohemoslovaca*, **20**, 57.

Chudakova, I. V. & Bocharova-Messner, O. M. (1965). Changes in the functional and structural features of the wing muscles of the domestic cricket (*Acheta domesticus* L.). *Doklady Akademii Nauk SSSR*, **164**, 656–9.

Crossman, A. R., Kerkut, G. A., Pitman, R. M. & Walker, R. J. (1971). Electrically excitable nerve cell bodies in the central ganglia of two insect species, *Periplaneta americana* and *Schistocerca gregaria*. Investigation of cell geometry and morphology by intracellular dye injection. *Comparative Biochemistry and Physiology*, **40**, 579–94.

Crossman, A. R., Kerkut, G. A. & Walker, R. J. (1971). Axon pathways of electrically excitable nerve cell bodies in the insect central nervous system. *Journal of Physiology*, **218**, 55–6.

Dingle, H. (1965). The relationship between age and flight activity in the milkweed bug, *Oncopeltus*. *Journal of Experimental Biology*, **42**, 269–83.

Edwards, J. S. (1969). Postembryonic development and regeneration of the insect nervous system. *Advances in Insect Physiology*, **6**, 97–137.

Edwards, J. S. & Palka, J. (1971). Neural regeneration: delayed formation of central contacts by insect sensory cells. *Science*, **172**, 591–4.

Farnworth, E. G. (1972). Effects of ambient temperature, humidity, and age on wing-beat frequency of *Periplaneta* species. *Journal of Insect Physiology*, **18**, 827–39.

Grellet, P. (1961). Cinetique du développement embryonnaire des Gryllides. *Bulletin biologique de la France et de la Belgique*, **95**, 613–43.

(1968). Milieu de culture pour embryons de gryllides. *Journal of Insect Physiology*, **14**, 1735–61.

Guthrie, D. M. (1964). Observations on the nervous system of the flight apparatus in the locust *Schistocerca gregaria*. *Quarterly Journal of Microscopical Science*, **105**, 183–201.

Gymer, A. & Edwards, J. S. (1967). The development of the insect nervous system. I. An analysis of postembryonic growth in the terminal ganglion of *Acheta domesticus*. *Journal of Morphology*, **123**, 191–7.

Heinig, S. (1967). Die Abanderung embryonaler Differenzierungsprozesse durch totale Rontgenbestrahlung im Ei von *Gryllus domesticus*. *Zoologische Jarhbucher. Abteilung für Anatomie und Ontogenie der Tiere*, **84**, 425–92.

Hinton, H. E. (1959). How the indirect flight muscles of insects grow. *Science Progress, London*, **47**, 321–33.

Hoyle, G. (1970). Cellular mechanisms underlying behavior–neuroethology. *Advances in Insect Physiology*, **7**, 349–444.

Huber, F. (1960). Untersuchungen uber die Funktion des Zentralnervensystems und insbesondere des Gehirnes bei der Fortbewegung und der Lauterzeugung der Grillen. *Zeitschrift für vergleichende Physiologie*, **44**, 60–132.

Ikeda, K. & Kaplan, W. D. (1970). Unilaterally patterned neural activity of gynandromorphs, mosaic for a neurological mutant of *Drosophila melanogaster*. *Proceedings of the National Academy of Science*, **67**, 1480–7.

Iles, J. F. & Mulloney, B. (1971). Procion yellow staining of cockroach motor neurones without the use of microelectrodes. *Brain Research*, **30**, 397–400.

Jobin, L. J. & Huot, L. (1966). Aperçu du développement embryonnaire et postembryonnaire du grillon domestique, *Acheta domesticus* (L). *Naturaliste Canadien*, **93**, 701–18.

Johansson, A. S. (1957). The nervous system of the milkweed bug, *Oncopeltus fasciatus* (Dallas) (Heteroptera, Lygaeidae). *Transactions of the American Entomological Society*, **83**, 119–83.

Kanellis, A. (1952). Anlagenplan und Regulationserscheinungen in der Keimanlage des Eies von *Gryllus domesticus. Wilhelm Roux Archiv für Entwicklungsmechanik der Organismen*, **145**, 417–61.

Kutsch, W. (1969). Neuromuskulare Aktivitat bei verschiedenen Verhaltensweisen von drei Grillenarten. *Zeitschrift für vergleichende Physiologie*, **63**, 335–378.

(1971). The development of the flight pattern in the desert locust, *Schistocerca gregaria. Zeitschrift für vergleichende Physiologie*, **74**, 156–68.

Kutsch, W. & Usherwood, P. N. R. (1970). Studies of the innervation and electrical activity of flight muscles in the locust *Schistocerca gregaria. Journal of Experimental Biology*, **52**, 299–312.

Levenbrook, L. & Williams, C. M. (1956). Mitochondria in the flight muscles of insects. III. Mitochondrial cytochrome C in relation to the aging and wingbeat frequency of flies. *Journal of General Physiology*, **39**, 497–512.

Levine, J. D. (1972). Neural control of flight in wild-type and mutant *Drosophila melanogaster*. Ph.D. Thesis, Yale University.

Mahr, E. (1961). Bewegungssysteme in der Embryonalentwicklung von *Gryllus domesticus. Wilhelm Roux Archiv für Entwicklungsmechanik der Organismen*, **152**, 662–724.

Michel, R. (1971). Variations des capacités de reponse 'vol continu' exprimees par le Criquet Pèlerin *Schistocerca gregaria* (Forsk) en fonction de l'âge du premier test. *Comptes rendus de l'Academie des sciences, Paris*, D, **272**, 1297–300.

Möss, D. (1971) Sinnesorgane im Bereich des Flügels der Feldgrille (*Gryllus campestris* L.) und ihre Bedeutung fur die Kontrolle der Singbewegung und die Einstellung der Flugellage. *Zeitschrifte für vergleichende Physiologie*, **73**, 53–83.

Mueller, N. S. (1963). An experimental analysis of molting in embryos of *Melanoplus differentialis. Developmental Biology*, **8**, 222–40.

Neville, A. C. (1963). Motor unit distribution of the dorsal longitudinal flight muscles in locusts. *Journal of Experimental Biology*, **40**, 123–36.

Nuesch, H. (1968). The role of the nervous system in insect morphogenesis and regeneration. *Annual Review of Entomology*, **13**, 27–44.

Panov, A. A. (1966). Correlations in the ontogenetic development of the central nervous system in the house cricket *Gryllus domesticus* L. and the mole cricket *Gryllotalpa gryllotalpa* L. (Orthoptera: Grylloidae). *Entomological Review*, **45**, 179–85.

Pitman, R. M., Tweedle, C. D. & Cohen, M. J. (1972). Branching of central neurons: intracellular cobalt injection for light and electron microscopy. *Science*, **176**, 412–14.

Rakshpal, R. (1962). Morphogenesis and embryonic membranes of *Gryllus assimilis* (Fabricius) (Orthoptera: Gryllidae). *Proceedings of the Royal Entomological Society of London*, A, **37**, 1–12.

Roderieck, R. W., Kiang, N. Y.-S. & Gerstein, G. L. (1962). Some quantitative methods for the study of spontaneous activity of single neurons. *Biophysical Journal*, **2**, 351–68.

Roth, L. M. (1948). A study of mosquito behavior. *American Midland Naturalist*, **40**, 265–352.

Sbrenna, G. (1971). Postembryonic growth of the ventral nerve cord in *Schistocerca gregaria* Forsk. (Orthoptera: Acrididae). *Bolletino di Zoologica*, **38**, 49–74.

Schneirla, T. C. (1948). Army-ant life and behavior under dry-season conditions with special reference to reproductive functions. II. The appearance and fate of the males. *Zoologica* (*New York*), **33**, 89–112.

Scudder, G. G. E. & Hewson, R. J. (1971). The postembryonic development of the indirect flight muscles in *Oncopeltus fasciatus* (Dallas) (Hemiptera: Lygaeidae). *Canadian Journal of Zoology*, **49**, 1377–86.

Svidersky, V. L. (1969). Receptors of the forehead of the locust, *Locusta migratoria* in ontogenesis. *Jurnal Evol'uzionnoj Biochimii i Fisiologii*, **5**, 482–90.

Teutsch-Felber, D. (1970). Experimentelle und histologische Untersuchungen an der Thoraxmuskulatur von *Periplaneta americana* L. *Revue Suisse de Zoologie*, **77**, 481–523.

Thomas, G. J. (1954). The post-embryonic development of the flight muscles of *Lamarckiana* sp. (Orthoptera) and a brief comparison of these with those of *Saussurea stuhlmanniana* (Karsch) and *Tanita dispar* (Miller). *Proceedings of the Royal Entomological Society of London*, A, 23–30.

Tiegs, O. W. (1955). The flight muscles of insects – their anatomy and histology; with some observations on the structure of striated muscle in general. *Philosophical Transactions of the Royal Society of London*, B, **238**, 221–347.

Tyrer, N. M. (1969). Time course of contraction and relaxation in embryonic locust muscle. *Nature, London*, **224**, 815–17.

Vogell, W. (1965). Phasen der Bildung morphologischer und enzymatischer Muster des Flugmuskels der Wanderheuschrecke. *Naturwissenschaften*, **52**, 405–18.

Voss, F. (1911). Uber den Thorax von *Gryllus domesticus*. Die nachembryonale Metamorphose im ersten Stadium. (Abdomen–metathorax). *Zeitschrift für Wissenschaftliche Zoologie*, **100**, 589–834.

(1912). Uber den Thorax von *Gryllus domesticus*. Die nachembryonale Metamorphose im ersten Stadium. (Mesothorax–head). *Zeitschrift für Wissenschaftliche Zoologie*, **101**, 445–682.

Walker, P. R., Hill, L. & Bailey, E. (1970). Feeding activity, respiration, and lipid and carbohydrate content of the male desert locust during adult development. *Journal of Insect Physiology*, **16**, 1001–15.

Warren, R. H. & Porter, K. R. (1969). An electron microscope study of differentiation of the molting muscles of *Rhodnius prolixus*. *American Journal of Anatomy*, **124**, 1–30.

Weber, T. (1972). Stabilisierung des Flugrhythmus durch 'Erfahrung' bei der Feldgrille. *Naturwissenschaften*, **59**, 366.

Wiesend, P. (1957). Die postembryonale Entwicklung der Thoraxmuskulatur bei einigen Feldheuschrecken mit besonderer Berucksichtigung der Flugmuskeln. *Zeitschrift für Morphologie und Okologie der Tiere*, **46**, 529–70.

Wigglesworth, V. B. (1953). The origin of sensory neurones in an insect, *Rhodnius prolixus* (Hemiptera). *Quarterly Journal of Microscopical Science*, **94**, 93–112.

(1956). Formation and involution of striated muscle fibres during the growth and moulting cycles of *Rhodnius prolixus* (Hemiptera). *Quarterly Journal of Microscopical Science*, **97**, 465–80.

Wilson, D. M. (1968). The nervous control of insect flight and related behavior. *Advances in Insect Physiology*, **5**, 289–338.

Notes added in proof

1. The material quoted as Kutsch (personal communication) has now been published (Kutsch, 1973).
2. With a new technique (low temperature axonal iontophoresis) some metathoracic dorso-medial neurons (pp. 166, 171; Fig. 7) have filled after the imaginal moult.
3. The midline ventral motor neuron cell bodies (p. 166; Fig. 7) may be inhibitory neurons (see Burrows, 1973).

Burrows, M. (1973). Physiological and morphological properties of the metathoracic common inhibitory neuron of the locust. *Journal of Comparative Physiology*, **82**, 59–78.

Kutsch, W. (1973). The influence of age and culture temperature on the wing-beat frequency of the migratory locust, *Locusta migratoria*. *Journal of Insect Physiology*, **19**, 763–72.

Specificity and regeneration in insect motor neurons

D. Young

Introduction

In order to move about, or indeed to remain absolutely still, an animal depends on the co-ordinated activity of its limb muscles. The muscles are activated by motor neurons located in the central nervous system (CNS) and the necessary connections between motor neurons and muscles are established during the growth of the animal. For the different muscles to interact in a co-ordinated manner there must be some source of specificity in the connections formed between nerve and muscle. This question of neuromuscular specificity has been investigated for several years, especially using the amphibia as experimental animals. This valuable line of work has been the subject of several recent reviews from various points of view (Gaze, 1970; Hughes, 1968; Jacobson, 1970; Mark, 1969).

Neuromuscular specificity has been little studied in insects by comparison with vertebrates. However recently it has become apparent that insects are very suitable for this kind of work and may in fact offer certain advantages. The insect leg is supplied by relatively few muscles, which are innervated by a comparatively small number of motor neurons. It is now possible to identify individually each of the motor neurons supplying a particular group of leg muscles. This offers the possibility of studying neuromuscular specificity using single identified neurons and their muscles, and so introduces a new level of precision which is not possible, as yet, in the vertebrates.

In hemimetabolous insects, whose postembryonic growth consists of several instars gradually approaching the adult condition, a number of species will regenerate lost parts and regrow nerve–muscle connections after these have been severed. In the present context, this means that these insects will accept the same range of regeneration and transplantation experiments as the amphibia, and these experiments may be used as a tool to pose questions about the sort of distinctions which motor neurons are capable of making during regrowth.

The cockroaches have proved to be the most suitable insects for this work and reference will be made mainly to *Periplaneta americana*, referring to other insects where appropriate. This chapter offers a brief review of that literature which is directly relevant to neuromuscular specificity in insects and a critical assessment of what has been achieved so far.

Insect motor innervation and its regeneration

The pattern of motor innervation

There is now a reasonable body of basic information about the insect neuromuscular system, which serves as an essential background for studies on specificity. Useful compilations are available for insects in general in Bullock & Horridge (1965) and for the cockroach in Guthrie & Tindall (1968). Reference should also be made to the useful recent reviews by Hoyle (1970) and Kandel & Kupfermann (1970). The following brief remarks are intended mainly to orientate those unfamiliar with the insect nervous system.

The three pairs of legs of the cockroach are innervated by three paired thoracic ganglia, which are fused laterally and joined longitudinally by paired connectives. Studies on motor physiology and regeneration have dealt mainly with the second thoracic (mesothoracic) and third thoracic (metathoracic) segments. The thoracic ganglia each give rise to eight segmental nerves, four of which supply the leg and associated muscles. In the cockroach, detailed descriptions and numbering are available for the nerve branches and their destinations (Dresden & Nijenhuis, 1958; Nijenhuis & Dresden, 1955; Pipa & Cook, 1959) and for the muscles (Carbonell, 1947; Dresden & Nijenhuis, 1953). The terminology of these authors is now standard and Guthrie & Tindall (1968) should be consulted for details.

The motor neuron cell bodies form a rind next to the ganglion sheath and distinct from the neuropile. Most are situated on the ventral and lateral surfaces but a few occur dorsally. The morphology of individual insect motor neurons has been known since the classical studies of Zawarzin, who showed that the cell body sends a slender process into the neuropile, where most of the dendrites spring from one branched region and the axon passes out along the peripheral nerve trunk (Zawarzin, 1924; see also Wigglesworth, 1959).

(*a*)

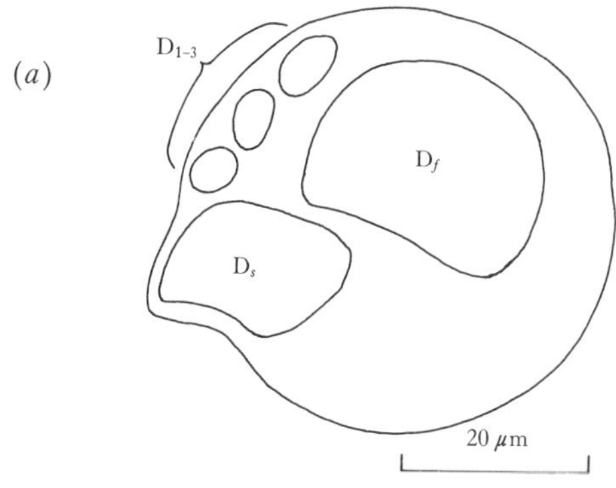

(*b*)

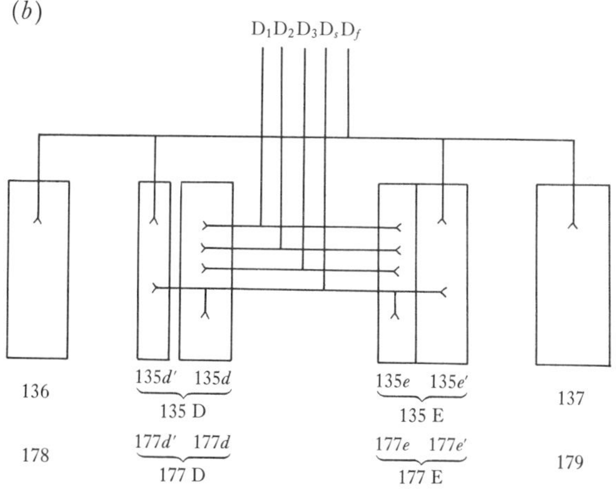

Fig. 1. Innervation of the coxal depressor muscles of the cockroach. (*a*) Transverse section through nerve 5r1 showing the outlines of the five motor axons; the rest of the nerve branch is filled with sensory axons of 4 μm or smaller. (*b*) Distribution of the five motor axons to the mesothoracic (135–7) and metathoracic (177–9) coxal depressor muscles. D_f, fast coxal depressor neuron; D_s, slow coxal depressor neuron; D_1, D_2, D_3, branches of three common inhibitory neurons. (From Pearson & Iles, 1971.)

The physiology of the motor supply was first studied in detail by Pringle (1939) who distinguished fast and slow fibres supplying the leg muscles. Although this distinction no longer holds for many arthropod muscles, it remains valid for the limb muscles of the cockroach (Pearson & Iles, 1970, 1971; Iles & Pearson, 1971). Fast neurons fire at low frequency, eliciting excitatory junction potentials (EJP) and powerful twitch contractions in the muscle fibres. These responses show antifacilitation. Slow neurons fire in high frequency bursts, eliciting EJPs and strong graded contractions, showing facilitation. In addition, peripheral inhibitory motor neurons have been demonstrated recently in insects (Pearson & Bergman, 1969; Usherwood, 1968). These elicit inhibitory (hyperpolarising) junction potentials and a reduction in the tension developed by the slow neurons. In the insect leg these inhibitory neurons innervate several muscle groups, and so are termed common inhibitory neurons. Inhibitory neurons supplying only one muscle or muscle group (specific inhibitory neurons) have not yet been reported in insects though they occur widely in the Crustacea.

The coxal depressor muscles of the cockroach make a good example of insect leg motor innervation and, since they will be discussed below, an outline of their innervation is given here. Fig. 1 shows in summary the innervation of the depressor muscles intrinsic to the coxa. Mesothoracic muscles 135D and E insert on a common apodeme and are each divided into two distinct parts, 135*d* and *d*′, 135*e* and *e*′. Muscles 136, 137 each have separate apodemes and lie respectively dorsal and ventral to muscle 135. The motor axons are identified by their electrical characteristics and correlated with the axons visible histologically in nerve 5*r*1 (Fig. 1*a*). The fast neuron, D_f, is the only neuron supplying muscles 136, 137 and also supplies some which receive slow innervation (135*d*′, *e*′). The slow neuron, D_s, supplies these muscles and neighbouring ones sharing inhibitory innervation (135*d*, *e*). The three inhibitory neurons, D_1–D_3, innervate muscles supplied by D_s but not those supplied by D_f (Fig. 1*b*).

Therefore in the context of specificity and regeneration, this work shows that the peripheral unit of study comprises a large group of muscle fibres having a particular anatomical location and distinct physiological characteristics. The central unit of study is the motor neuron. Specificity then refers to the constancy of these units and of the connections between them.

Regeneration of motor nerves

For a general review of the interdependence of nerve and muscle in insects reference should be made to Nuesch (1968). An experimental study by Teutsch-Felber (1970) of postembryonic muscle growth in the cockroach also provides material relevant to the study of motor regeneration. In a study of the regeneration of lost limbs, Bodenstein (1955) showed that cockroach nerves possess a considerable power of regeneration and that this ability is not confined to the juvenile stages but is retained by the nerves of the adult animal. In fact he could not obtain a regenerate without innervation because even if the entire ganglion was removed from a segment, fibres grew out of the cut connectives to re-innervate the regenerating leg stump. In a further study, Bodenstein (1957) showed that following a less drastic operation, such as severance of the main leg nerve (N5) without removing the leg, the nerve can repair itself and the motor neurons restore functional contact with the muscles.

Guthrie (1962, 1967) followed up this work with a study of regeneration of the innervation of the femoral muscles and Jacklet & Cohen (1967*b*) repeated Bodenstein's experiments on regeneration of nerve 5, recording with microelectrodes from the muscles. Taking the results of all these authors in combination, the repair of nerve 5 can be seen to consist of the following sequence of events. By 5–6 days after the nerve has been severed the muscles have become electrically inexcitable, miniature end-plate potentials can no longer be recorded and the distal nerve stump and terminals show histological changes indicative of degeneration. At about 10–11 days the first fibres can be observed bridging the gap between the cut ends of nerve 5. In nymphs, motor neuron contact with the muscles is restored from about 14 days onwards. In the adult, miniature end-plate potentials and muscle potentials appear from about 30 days onwards. Although Bodenstein gives a more extended figure for nymphs, overall comparison between these authors indicates that the adults regenerate rather less rapidly and less readily than do nymphs. At first the electrical responses in regenerates are not normal in amplitude and frequency, but with time they approixmate to the normal condition.

Nerve regeneration has been little studied in other insects. Crickets are certainly capable of regenerating lost limbs (E. Ball, personal communication) and cerci (Edwards, this volume) and will re-innervate transplanted limbs (Sahota & Edwards, 1969). Hence their neural regeneration is probably as vigorous as that of the cockroach but it has not been studied systematically. Locusts are more doubtful. The response of their adult muscles to denervation has been studied (Usherwood, 1963*a*, *b*) and is similar to the cockroach except for the time scale, but there is no adequate published evidence that their motor neurons are capable of regenerating to restore functional contact with the muscles. There is some evidence for the regeneration of central neurons in locusts (Boulton, 1969; Boulton & Rowell, 1969) but this is rather inconclusive and has yet to be confirmed by others. So far, then, the cockroach is the most suitable insect for experimental studies on neuromuscular specificity because it is known that its nerves can regenerate vigorously and because the anatomy and physiology of its neuromuscular system is known in reasonable detail.

Recent techniques of study

There are a number of methods of studying motor neurons, some of which have only become available quite recently, which are specially applicable to the study of their specificity. The histological arrangement of the larger neuron cell bodies can be shown by means of projection reconstruction from serial sections using the method of Pusey (1939) and with experience it is advantageous to use the rapid modification of Potts (1966). I have always employed 10 μm transverse sections and projected them to obtain a ventral view. This enables one to form a picture of the three-dimensional arrangement of the cell bodies fairly readily and with a little practice one can reconstruct mentally as one scans a series of transverse sections.

When these methods of reconstruction are combined with a study of the cell-body response to axonal injury (Cohen & Jacklet, 1965) they can be used to produce cell maps showing which cell bodies send their axons down which nerve trunk (Cohen & Jacklet, 1967; Young, 1969). Experience with this method shows that the injury response is apparent only after heavy injury to the motor neuron and cannot be discerned if the axon is cut more than a millimetre or two distal to the ganglion. Consequently, despite early optimism, this method cannot be used to locate cell bodies of neurons innervating individual limb muscles. For such study this method has been superseded by others described below but the injury response itself deserves further study. The most obvious way of accounting for the chromatolytic reaction of the injured cells observed in the light microscope would be by equivalent ultrastructural changes in endoplasmic reticulum and ribosomes (Cohen, 1967), but an electron microscope study did not confirm this (Young, Ashhurst & Cohen, 1970). This would be worth a fresh study with the more refined biochemical methods currently available.

Following the introduction of the dye procion yellow by Stretton & Kravitz (1968) it has become possible to mark individual neurons which have been penetrated with a microelectrode. Bentley (1970), Burrows & Hoyle (1973) and Hoyle & Burrows (1973) have shown that insect motor neurons can be triggered by intracellular stimulation of the cell body and the muscle response observed. The neuron is then marked by electrophoretically injecting the cell body with procion yellow from the microelectrode. This makes it possible to locate the cell bodies of neurons supplying particular muscles. Since recordings can also be made with dye-filled electrodes penetrating the cell body, the activity of the identified neurons can be monitored. Careful filling with dye also makes it possible to study the dendritic field of identified motor neurons (Davis, 1970; Sandeman, 1971). A superior method of doing this has been published by Pitman, Tweedle & Cohen (1972) employing cobalt chloride which is electrophoretically injected into the cell from the microelectrode and subsequently rendered visible by reaction with ammonium sulphide. The cobalt penetrates the finest dendritic branches relatively quickly and the marked cell can be studied in fixed and cleared whole mounts without the need for serial sectioning. In addition Iles & Mulloney (1971) devised a method for electrophoretically passing procion yellow into the cell through the cut end of the peripheral axon. In this way dye will travel several millimetres and can be used to mark groups of neurons innervating a given peripheral site. This method produces considerable swelling with procion but the same method works equally well with cobalt without any swelling effect.

The usefulness of extracellular recording techniques should not be overlooked. When recording from the peripheral nerve branches close to the muscles, several motor axons can be distinguished by the amplitude and other characteristics of their action potentials and these can be correlated with histological sections of the peripheral nerve branch (Pearson, Stein & Malhotra, 1970; see also Fig. 1, above). Thus once an identified neuron has been characterised in this way, its activity can be monitored peripherally without microelectrode recording.

With this considerable arsenal of recently available techniques, one may expect that the physiological study of identified motor neurons will increase exponentially. At the same time, these techniques provide the opportunity of studying the specificity of connections of identified motor neurons in a suitable animal such as the cockroach and this opportunity has not been available in earlier studies on neuromuscular specificity.

Identity of motor neurons

Cohen & Jacklet (1967) counted a total of about 3000 neuron cell bodies in the cockroach metathoracic ganglion. About 200 of these are 20 μm or larger in diameter and these are included in the cell maps reconstructed by the technique described above. Just over half these larger neurons can be recognised as occurring consistently in bilaterally symmetrical pairs and identified from one animal to another. I originally devised the numbering system for these cell bodies while an assistant on the project that resulted in Cohen & Jacklet's cell map of the metathoracic ganglion and I subsequently extended it to the mesothoracic ganglion (Young, 1969). Each of the larger, bilaterally symmetrical cell bodies is assigned a number on the basis of its size, shape and position, proceeding in an arbitrary way from anterior to posterior (Fig. 2). The mesothoracic and metathoracic ganglia of the cockroach are very similar and serially homologous cell bodies can be recognised in each. By this is meant only that the two ganglia are so similar that it is possible to recognise histologically 'the same' cell body in the two serial, segmental ganglia. Thus mesothoracic cell bodies 29 and 30 correspond to metathoracic cell bodies 27 and 28 (Fig. 3). Such morphological judgements of homology are not dependent on correlation with other cell characteristics. For instance, if homologous cell bodies were shown not to innervate homologous muscles, this would not mean that the cell bodies were not homologous after all, but only that homology of cell body characteristics and of axonal destination were independent variables.

The characteristics by which cell bodies are recognised are variable but the more conspicuous cell bodies can still be recognised even when one of the usual characteristics is anomalous. Hence 'errors' in cell-body position or other features can be detected.

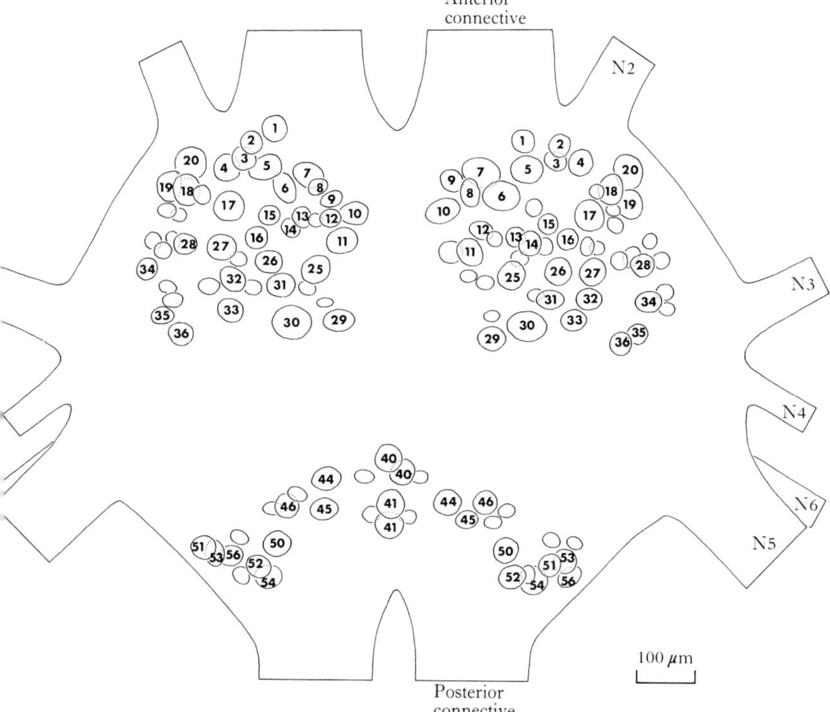

Fig. 2. Cell map of the mesothoracic ganglion of the cockroach, showing the distribution of neuron cell bodies over 20 μm in diameter. Reconstructed from serial 10 μm sections, the ganglion is viewed from the ventral aspect. Numbers are allotted to the larger, bilaterally symmetrical neuron cell bodies. N2–N6, peripheral nerve trunks.

The sensory neurons of the cockroach leg can also be recognised individually on the basis of their size, shape and position (Young, 1970). Another similar example of individual cell recognition is the leech ganglion, where single motor and sensory neurons can be identified by their electrical and morphological features (Nicholls & Baylor, 1968; Stuart, 1970) and these neurons are being used in regeneration studies.

The cockroach has not yet been subjected to a systematic microelectrode study to determine which cell bodies belong to motor neurons supplying individual muscles, though a few have been so identified (Iles, 1972; Young, 1972). But such a study has

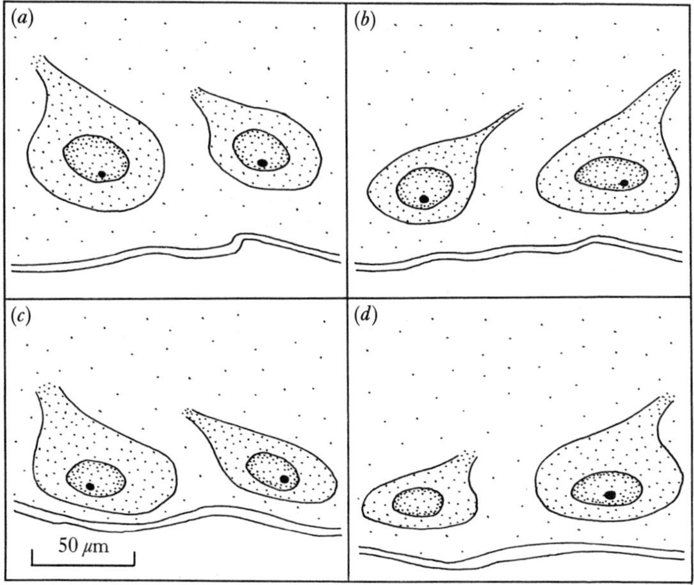

Fig. 3. Drawings of identified cockroach thoracic motor neurons seen in transverse sections to show the similarity between homologous cell bodies on either side of the same ganglion and between ganglia. *Above*, cell bodies 29 and 30 from (*a*) the left-hand side and (*b*) the right-hand side of the mesothoracic ganglion. *Below*, cell bodies 27 and 28 from (*c*) the left-hand side and (*d*) the right-hand side of the metathoracic ganglion. (Modified from Young, 1969.)

been carried out on the locust (Bentley, 1970; Burrows & Hoyle, 1973; Hoyle & Burrows, 1973). These authors find that the neuron cell body supplying a given muscle is consistently located in the same region of the ganglion. Although the locust neurons are not identified on morphological criteria, their results taken in conjunction with the cockroach examples strongly suggest that the same identified cell body will always be found to innervate the same muscle. Of course purely morphological identification of cell bodies is unimportant for physiological work but it is essential for studies of neuromuscular specificity. In order to carry out meaningful experiments on the specificity of the connections of motor neurons to their muscles it is necessary to identify individual motor neurons independently of their demonstrated connections to given muscles.

The cell maps of locust and cockroach show that the majority of the larger cell bodies in the ganglia belong to motor neurons. Their layout in the two insects is broadly similar. The majority of excitatory leg motor neurons are situated in the anterior part of the ganglion but some, particularly an important group of neurons supplying bifunctional leg/flight muscles, are situated posteriorly. The highly consistent positioning of the cell bodies shows that their arrangement is not random, but there is no simple correlation of cell body position with axonal destination. A comparison suggests that similar factors may determine cell-body location in both locust and cockroach. One factor may be the tendency for cell bodies of fast neurons supplying functionally interdependent muscles to be located close together in the way described for the crab eye-withdrawal muscles (Burrows & Horridge, 1968; Sandeman, 1971). Thus, in the cockroach, the conspicuous pair of mesothoracic neurons, cells 29 and 30, supply functionally interdependent muscles concerned with fast depression of the leg (Fig. 5; Young, 1972). In this ganglion, cell 33, located close to cell 30 (Fig. 2), supplies the fast retractor unguis (Young, unpublished results). In the locust the fast retractor unguis cell body is located close to the fast extensor tibiae (Burrows & Hoyle, 1973). Thus in both insects the fast cell keeping the tarsus firmly adpressed to the ground is located close to the fast cell delivering the main locomotory thrust to the leg. However, the slow neuron cell bodies to the cockroach coxal depressors and the locust extensor tibiae are each rather widely separated from the cell bodies of the fast neurons to these muscles (Burrows & Hoyle, 1973; Iles, 1972).

The common inhibitory neurons of the locust have their bilaterally symmetrical cell bodies close together in a rather distinct group of cell bodies situated posteriorly (Burrows, 1973). A similarly distinct group of cell bodies occurs in the cockroach ganglia, which I have numbered from 40–6 in Fig. 2. These cells were left without numbers in Cohen & Jacklet (1967) and Young (1969), because injury responses did not show their association with any one nerve trunk. Pearson (personal communication) has recorded from a cell body in this group which follows a peripherally recorded common inhibitory neuron in a 1 to 1 manner, and so it is very probable that the 40 group of cell bodies in the cockroach includes the common inhibitory cell bodies as in the locust. Burrows (1973) has suggested that their location together

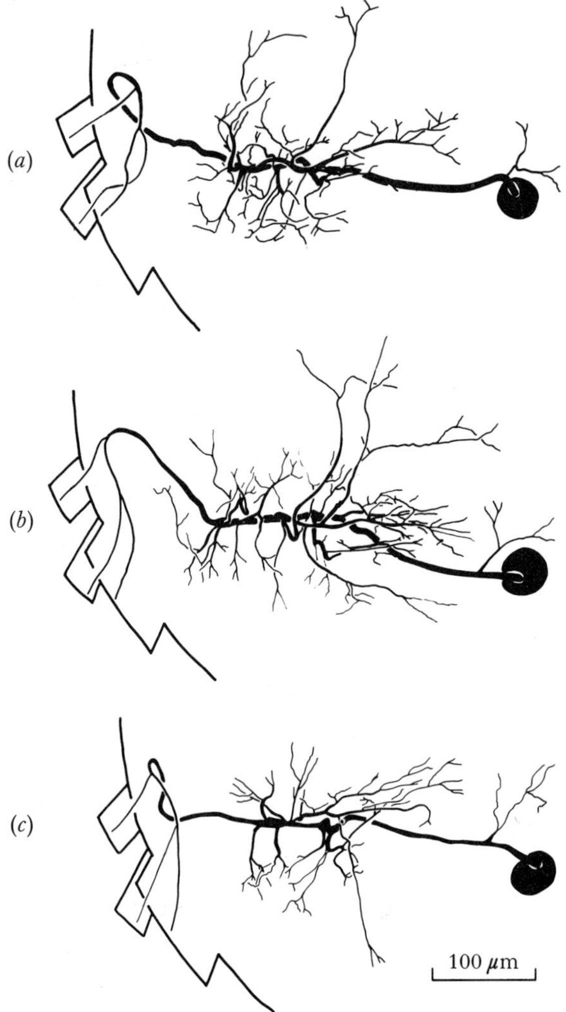

Fig. 4. (*a*) (*b*) (*c*) Three examples from different animals of the same common inhibitory neuron in the metathoracic ganglion of the locust. The cells have been filled with cobalt chloride via a microelectrode penetrating the cell body. Comparison of the three shows the degree of similarity and variation in the detailed structure of an identified motor neuron. (From Burrows, 1973. Reproduced by permission of Springer-Verlag, Berlin, Heidelberg & New York.)

in a distinct group might be on account of their sharing a common, distinct transmitter substance, a mode of grouping well known in other invertebrate ganglia (see review of Kandel & Kupferman, 1970). Thus although there are indications of functional factors governing the highly specific cell-body location, it is evident that no one simplistic functional theory will apply to all. It is still too early to write a paper on the functional organisation of motor neurons in an insect ganglion.

The dendritic patterns revealed by cobalt injection give a rather more sensitive indication of the structural individuality of insect motor neurons than is possible with cell-body characteristics. Since this field of study will now expand very rapidly, it is premature to generalise. But the first examples of fillings of identified motor neurons provide some valuable pointers. Fig. 4 is taken from Burrows (1973) and shows three fillings of the same common inhibitory neuron of the locust. The general pattern of branching is consistent and characteristic enough to permit identification of the neuron. But there are also several differences even at the level of first order branches. While some of these may be due to variations in histological procedure, many cannot be explained in this way and so reflect genuine variation in neuron structure.

The results described in this section all combine to show that single insect motor neurons can be identified from one animal to another and are characterised by a high degree of individuality. The various features, cell body characteristics, electrical characteristics, dendritic branching etc., though variable, are all highly specific for a given neuron.

Regeneration in leg motor neurons

The work of Pearson and Iles referred to above shows how the very constant pattern of innervation of the coxal depressor muscles of the cockroach relates to their normal mode of functioning. This situation lends itself well to the detailed study of the specificity of motor neuron regeneration following severance of the leg nerve. That approximately normal functioning can be restored is clear from the earlier work on regeneration following severance of nerve 5, particularly that of Guthrie (1967). He reports that the restoration of function is never perfect but is frequently very good, especially in nymphs. To restore such normal functioning it is necessary not only that the depressor muscles receive depressor, rather than levator re-innervation but also that within the small group of coxal depressor muscles each receives appropriate physiological innervation. At some point in the system the restored connections must be physiologically as well as anatomically correct. There are two obvious ways of achieving this; either by specific regrowth of the correct motor neurons or by random regrowth of the motor neurons followed by myotypic modulation of their central connections.

This point is being studied by Pearson & Bradley (1972) using the coxal depressor motor neurons of the cockroach metathoracic segment. Muscle recordings in freely walking animals (technique of Pearson, 1972) shows that after regeneration the coxal depressor muscles are re-innervated by two excitatory motor neurons which have discharge patterns similar to D_f and D_s of normal animals. Recordings from nerve 5*r*1 in regenerate animals have shown that this nerve contains spontaneously active motor axons with discharge patterns similar to axons D_1, D_2, D_3 and D_s in normal animals (axon D_f never fires spontaneously). Intracellular muscle recording in regenerate animals shows that the distribution of the fast and slow axon is the same as normal, that is to say the fast neuron supplies 177*d*′, 177*e*′, and 178 and 179 but not 177*d* or 177*e*, while the slow neuron supplies all parts of 177d and 177e but not 178 or 179. Finally, the sizes and positions of the fast and slow cell bodies appear to be identical to D_f and D_s of normal animals, and intracellular recording and stimulation via these cell bodies confirms the peripheral identifications. This last point, with the morphological evidence of cell body location, is crucial here because we cannot know in advance to what extent myotypic specification, if it occurred, might be able to modulate the electrical characteristics of a randomly connected neuron to resemble the correct pattern of innervation. This evidence therefore supports the idea of specific regrowth by the identified motor neurons to their original target muscles and not to others. The coxal depressor muscles also receive inhibitory re-innervation in regenerates but it is not yet certain that it is the same as normal.

From here the way is opened to study some aspects of the mechanism of this specific regrowth, for which it will be necessary to study the time course of regeneration, as well as the end result. The earlier work on the regeneration of nerve 5 indicates that the distal stump of the nerve is kept intact by glial cells and regenerating sensory axons, and that the regrowing motor neurons find their way from the proximal to the distal stump. This may well guide the general direction of their regrowth, but several possibilities exist for control of the finer details. If the distal stumps of the individual motor axons do not completely degenerate, then matching could occur between proximal and distal stumps of single motor axons. However, Nordlander & Singer (1972) have presented evidence that this does not occur in Crustacea, where the distal axon stumps do not degenerate (but see Hoy, this volume, for fuller discussion and a different conclusion). If, as seems more likely, the distal stumps do degenerate then correct regrowth could be achieved either by specific biochemical matching between regrowing axons and their muscles or by random regrowth followed by specific rejection of incorrect synaptic connections. Guthrie (1967) states that in the early stages of restored contact with the muscles there are only a few widely scattered terminals which subsequently become more restricted and he describes how the electrical responses are abnormal at first but gradually improve. He suggests that slow neurons regrow first, followed by fast. This would be consistent with the idea of random regrowth followed by selective rejection, but it could also be compatible with specific regrowth. For, in an important contribution, Baylor & Nicholls (1971) and Jansen & Nicholls (1972) have provided evidence in the leech of a high degree of specificity in the regenerated connections between identified sensory and motor cells in the CNS, but the performance of the regenerated synapses is consistently altered with respect to the balance between excitation and inhibition. Hence it is possible that the improvement of electrical responses in regenerated neuromuscular connections could be due to improvement of performance in a correctly matched synapse, as well as to random regrowth followed by specific rejection. If the latter is the correct interpretation one should be able to find the supernumerary, rejected terminals by electron microscopy. These need not necessarily show degeneration for Mark, Marotte & Johnstone (1970) have shown in fish eye muscle that incorrect terminals persist structurally but are switched off functionally following correct re-innervation. These considerations serve to show that there is a whole set of related questions which invite study at the single cell level using identified motor neurons and their muscles. And the cockroach work referred to here provides the sort of framework within which such studies can be carried out.

Experimental manipulation of motor neuron regeneration

As well as studies in which a cut nerve is left free to regenerate, other studies involve experimental manipulation of the regenerating situation. In this way, regenerating motor neurons can be faced with an unusual situation and so be made to reveal their capacities for regrowth under these altered conditions. Since the motor supply is located in discrete segmental ganglia, one possibility is to transplant a ganglion into a new situation. Bodenstein (1957) transplanted a thoracic ganglion into the coxa in nymphal cockroaches, where he cut away some muscle to make room for it. When these animals became adult, he was able to show that such a ganglion innervated the host's coxal muscles by cutting the host ganglion's peripheral nerves and still being able to obtain myograms from the coxal muscles 16 days afterwards, by which time the host's motor axons definitely would have degenerated. Jacklet & Cohen (1967*a*) carefully repeated Bodenstein's experiments and provided evidence that the motor axons of the transplanted ganglion preferentially innervated those muscles which had been denervated by the transplant operation. Their rather stronger statement that only denervated muscles will accept regenerating fibres should be treated with caution, since only five animals were studied and since evidence for hyperinnervation is available from Crustacea (Hoy, this volume). Jacklet & Cohen (1967*a*) suggested that the ganglion transplant provides an *in-vivo* tissue culture preparation with the possibility of constructing an integrated excitable system where the connections can be specified, and this almost visionary optimism has been echoed by Jacobson (1970, p. 314). However, the histological appearance of the transplanted ganglion becomes considerably altered, thereby distorting the spatial distribution of cell bodies, and the overall number of cell bodies is often rather reduced. This makes cell body identification virtually impossible. Moreover, the transplanted ganglion remains surrounded by tough wound tissue for many months, which effectively prevents study with microelectrodes. These results from my own unpublished observations make it less likely that ganglion transplants will be a very useful tool in specificity studies. At all events, no further publications employing this technique have appeared.

Another possibility for manipulation of the nervous system is to cross motor nerves in a manner similar to the experiments that have been done on vertebrates. In an incomplete report, Cohen (1969) mentions experiments in which levator and depressor nerve trunks were crossed in the cockroach using plastic collars to guide the nerves. Unfortunately, these experiments are not reported in sufficient detail to enable them to be evaluated and in particular no details are given of any control measures designed to make sure that the regrowing nerves had not escaped from their constraints and returned to their normal destinations. Nevertheless, this nerve-crossing experiment is potentially a very valuable one and deserves to be repeated in detail with rigorous controls. Whatever the outcome, it should yield important information about the selective abilities of regrowing motor neurons.

An alternative approach to manipulating the motor nerves is to manipulate their target organs, the muscles. This is most readily accomplished by limb transplantation and I have carried out some experiments using this method (Young, 1972). In the cockroach, the mesothoracic and metathoracic limbs are similar in external appearance and in their musculature. The similarity of the mesothoracic and metathoracic ganglia made possible the recognition of serially homologous cell bodies in each on purely morphological criteria (see above). By penetrating such cell bodies with a dye-filled microelectrode and stimulating intracellularly, it was possible to show that such serially homologous cell bodies innervate serially homologous muscles in the two limbs (Fig. 5). Thus mesothoracic cell body 30 belongs to the mesothoracic D_f neuron and its homologue, metathoracic cell body 28, is the metathoracic D_f (Fig. 5; cf. Fig. 3). Following on from this, a metathoracic limb was transplanted on to the mesothoracic segment in a number of nymphal cockroaches. The transplanted limbs became incorporated in the mesothoracic segment but retained their metathoracic form. When the individuals with successful transplants became adult, the mesothoracic cell bodies were penetrated with a microelectrode and were found to have innervated those metathoracic muscles which are the serial homologues of their own target muscles. The

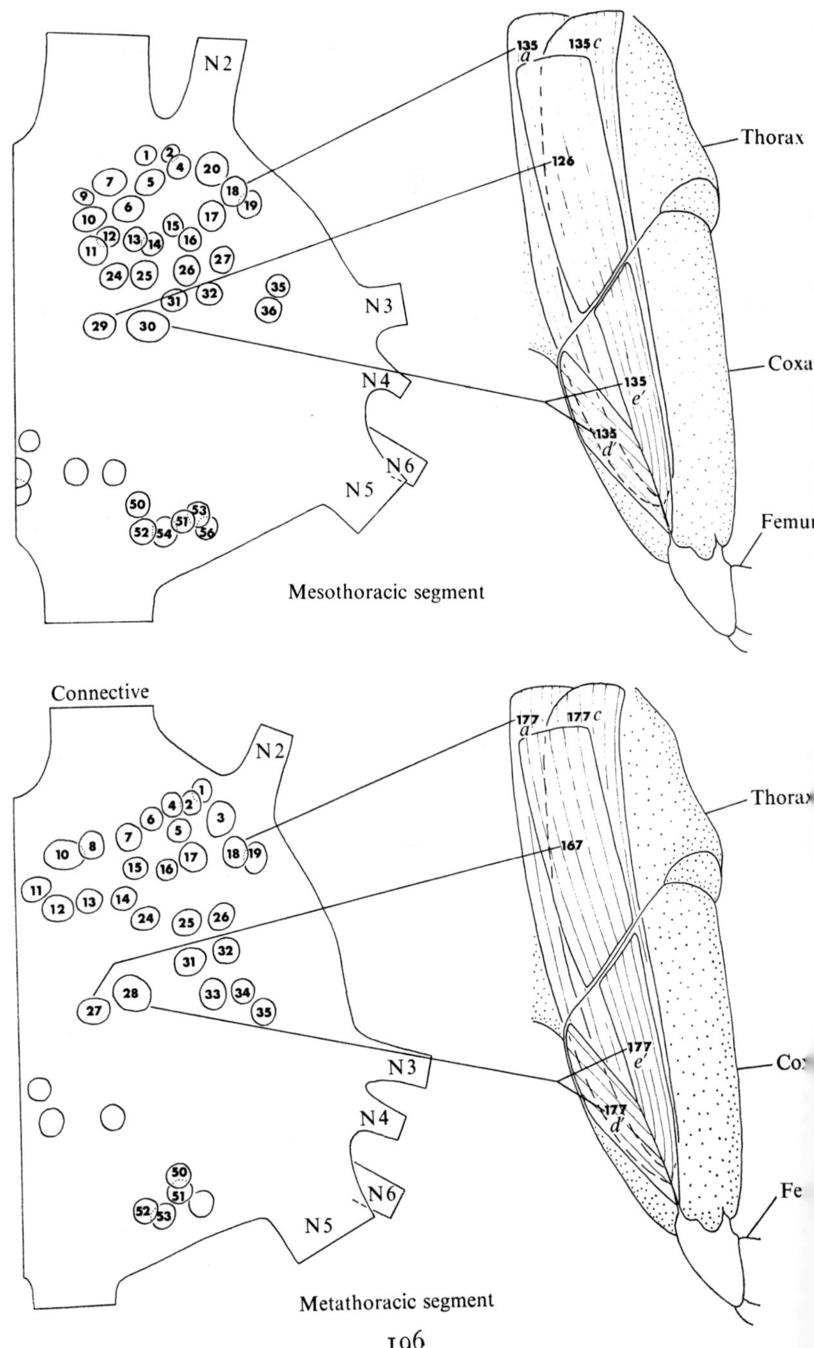
N 2
N 3
N 4
N 5
N 6
135 a
135 c
126
135 e′
135 d′
Thorax
Coxa
Femur
Mesothoracic segment
Connective
N 2
N 3
N 4
N 5
N 6
177 a
177 c
167
177 e′
177 d′
Thorax
Co
Fe
Metathoracic segment

transplanted limbs were capable of approximately normal walking movements.

This ability of mesothoracic motor neurons to innervate and distinguish between metathoracic muscles shows that the neuron–muscle specificity is not unique for each neuron and muscle. Rather there must be some way of matching them which is repeated segmentally. Evidentally there is a homology between segments in the nervous system as there is also in the cuticle, so that the mechanism of specification could be similar in both cases at least to some extent (cf. Bate & Lawrence, this volume). A further step along this line of work would be to repeat the same experiment on insects in which there are much greater differences in the musculature between segmental limbs as in crickets or mantids.

Sahota & Edwards (1969) successfully transplanted an additional leg on to the mesothoracic segment of the house cricket. The mesothoracic neurons were thus presented with an extra set of mesothoracic muscles to innervate. Some of the transplanted legs received innervation and appeared to make active, unco-ordinated movements but beyond this the results were inconclusive. As Sahota & Edwards themselves suggest, this experiment invites repetition using electrophysiological study of identified motor neurons. Such overloading of the motor system might be one situation in which motor neurons could be persuaded to make non-specific connections even in the presence of their correct target muscles.

Altogether there are a range of promising experimental manipulations of the motor system during regeneration. By means of a suitable combination of these, one should be able to determine what distinctions a motor neuron is capable of making during regrowth and under what conditions specificity of connection is either maintained or lost. This line of work would dovetail in with detailed studies of nerves allowed to regenerate freely as described

---

Fig. 5. Diagram illustrating the connections of selected identified motor neurons in the mesothoracic and metathoracic segments of the cockroach. For each segment is shown: left, an accurately reconstructed half-ganglion with identified motor neurons (numbered) and peripheral nerve trunks (N2–N6) and right, a drawing of the proximal parts of the leg with numbered muscles. The connecting lines join the selected cell bodies to the muscles which they supply. Muscles 136, 137 and 178, 179 have been omitted for clarity. (From Young, 1972.)

in the previous section. Taken together, these should make it possible to build up quite a detailed picture of the patterns of regrowth in identified motor neurons.

Conclusions

In the insects, single motor neurons can be identified by a variety of morphological and electrical features. A high degree of individuality is evident in these features and this enables the same neuron to be identified consistently from one animal to another. Though the motor neurons are few in number, the peripheral connections of these neurons can be quite complex and the central dendritic branching is very complex. This renders naive the idea that the insect motor system can be used as a simplified model of the vertebrate system. But the value of insect studies is not dependent on any simplified resemblance to the vertebrates because established differences between the various invertebrate and vertebrate mechanisms are just as likely to yield important insights for our understanding of neural specificity. The ability to work with single identified motor neurons makes the insects advantageous material for studies on neuromuscular specificity. As yet, this is a situation of promise for the future rather than of present achievement. The evidence so far obtained points to the capacity of insect motor neurons for specific regrowth, probably by means of some segmentally repeated matching system. At present, the insects can be exploited best by a series of experiments on regenerating motor neurons under a range of conditions. Each set of experiments should be carefully designed and rigorously controlled so as to pose, and preferably to answer, a specific question about the capabilities of regrowing neurons. In this way, one could compile an inventory of the distinctions which a regrowing motor neuron is or is not capable of making. From this it should be possible to infer some of the factors governing neuromuscular specificity during regeneration, or at least to infer the consequences of their expression. The interpretation of this work is limited by the fact that the neurons are regrowing in an otherwise preformed situation. Hence the means by which a neuron restores previously formed connections may not be the same as those which obtained in the initial, embryonic period of growth. But, bearing this in mind, the study of neuromuscular regrowth forms an important part of the overall study of neuronal specificity. The work reviewed in this chapter indicates that studies on insects have the potentiality of making a valuable contribution to this field.

I am specially indebted to Dr K. G. Pearson, Dr J. F. Iles and Dr M. Burrows for allowing me to see their unpublished material and for permission to use figures from their publications. Dr J. Altman, Dr E. Ball and Dr C. M. Bate gave helpful comments on the manuscript.

References

Baylor, D. A. & Nicholls, J. G. (1971). Patterns of regeneration between individual nerve cells in the central nervous system of the leech. *Nature, London*, **232**, 268–70.

Bentley, D. R. (1970). A topological map of the locust flight system motor neurons. *Journal of Insect Physiology*, **16**, 905–18.

Bodenstein, D. (1955). Contributions to the problem of regeneration in insects. *Journal of Experimental Zoology*, **129**, 209–24.

(1957). Studies on nerve regeneration in *Periplaneta americana*. *Journal of Experimental Zoology*, **136**, 89–116.

Boulton, P. S. (1969). Degeneration and regeneration in the insect central nervous system. I. *Zeitschrift für Zellforschung und Mikroskopische Anatomie*, **101**, 98–118.

Boulton, P. S. & Rowell, C. H. F. (1969). Degeneration and regeneration in the insect central nervous system. II. *Zeitschrift für Zellforschung und Mikroskopische Anatomie*, **101**, 119–34.

Bullock, T. H. & Horridge, G. A. (1965). *Structure and Function in the Nervous Systems of Invertebrates*, 2 vols. San Francisco & London: W. H. Freeman.

Burrows, M. (1973). Physiological and morphological properties of the metathoracic common inhibitory neuron of the locust. *Journal of Comparative Physiology*, **82**, 59–78.

Burrows, M. & Horridge, G. A. (1968). Eyecup withdrawal in the crab, *Carcinus*, and its interaction with the optokinetic response. *Journal of Experimental Biology*, **49**, 285–97.

Burrows, M. & Hoyle, G. (1973). Neural mechanisms underlying behavior in the locust *Schistocerca gregaria*. III. Topography of limb motoneurons in the metathoracic ganglion. *Journal of Neurobiology*, **4**, 167–86.

Carbonell, C. S. (1947). The thoracic muscles of the cockroach, *Periplaneta americana*. *Smithsonian Miscellaneous Collections*, **107**, no. 2.

Cohen, M. J. (1967). Correlations between structure, function and RNA metabolism in central neurons of insects. In *Invertebrate Nervous Systems*, ed. C. A. G. Wiersma, pp. 65–78. Chicago: University of Chicago Press.

(1969). Neuronal change in the regenerating and developing insect nervous system. In *Cellular Dynamics of the Neuron*, ed. S. H. Barondes, pp. 263–75. New York & London: Academic Press.

Cohen, M. J. & Jacklet, J. W. (1965). Neurons of insects: RNA changes during injury and regeneration. *Science*, **148**, 1237–9.

(1967). The functional organisation of motor neurons in an insect ganglion. *Philosophical Transactions of the Royal Society of London*, B, **252**, 561–72.

Davis, W. J. (1970). Motoneuron morphology and synaptic contacts: determination by intracellular dye injection. *Science*, **168**, 1358–60.

Dresden, D. & Nijenhuis, E. D. (1953). On the anatomy and mechanism of motion of the mesothoracic leg of *Periplaneta americana. Proceedings, Koniklijke Nederlandse Akademie van Wetenschappen*, **56**, 39–47.

(1958). Fibre analysis of the nerves of the second thoracic leg in *Periplaneta americana. Proceedings, Koniklijke Nederlandse Akademie van Wetenschappen*, **61**, 213–23.

Gaze, R. M. (1970). *The Formation of Nerve Connections*. London & New York: Academic Press.

Guthrie, D. M. (1962). Regenerative growth in insect nerve axons. *Journal of Insect Physiology*, **8**, 79–92.

(1967). The regeneration of motor axons in an insect. *Journal of Insect Physiology*, **13**, 1593–611.

Guthrie, D. M. & Tindall, A. R. (1968). *The Biology of the Cockroach*. London: Edward Arnold.

Hoyle, G. (1970). Cellular mechanisms underlying behavior – neuroethology. *Advances in Insect Physiology*, **7**, 349–444.

Hoyle, G. & Burrows, M. (1973). Neural mechanisms underlying behavior in the locust *Schistocerca gregaria*. I. Physiology of identified motoneurons in the metathoracic ganglion. *Journal of Neurobiology*, **4**, 3–42.

Hughes, A. F. W. (1968). *Apsects of Neural Ontogeny*. London: Logos Press.

Iles, J. F. (1972). Structure and synaptic activation of the fast coxal depressor motoneurone of the cockroach, *Periplaneta americana. Journal of Experimental Biology*, **56**, 647–56.

Iles, J. F. & Mulloney, B. (1971). Procion yellow staining of cockroach motor neurones without the use of microelectrodes. *Brain Research*, **30**, 397–400.

Iles, J. F. & Pearson, K. G. (1971). Coxal depressor muscles of the cockroach and the role of peripheral inhibition. *Journal of Experimental Biology*, **55**, 151–64.

Jacklet, J. W. & Cohen, M. J. (1967*a*). Synaptic connections between a transplanted insect ganglion and muscles of the host. *Science*, **156**, 1638–40.

(1967*b*). Nerve regeneration: correlation of electrical, histological, and behavioral events. *Science*, **156**, 1640–43.

Jacobson, M. (1970). *Developmental Neurobiology*. New York: Holt, Rinehart & Winston.

Jansen, J. K. S. & Nicholls, J. G. (1972). Regeneration and changes in synaptic connections between individual nerve cells in the central nervous system of the leech. *Proceedings of the National Academy of Sciences, USA*, **69**, 636–9.

Kandel, E. R. & Kupfermann, I. (1970). The functional organisation of invertebrate ganglia. *Annual Review of Physiology*, **32**, 193–258.

Mark, R. F. (1969). Matching muscles and motoneurones. A review of some experiments on motor nerve regeneration. *Brain Research*, **14**, 245–54.

Mark, R. F., Marotte, L. R. & Johnstone, J. R. (1970). Reinnervated eye muscles do not respond to impulses in foreign nerves. *Science*, **170**, 193–4.

Nicholls, J. G. & Baylor, D. A. (1968). Specific modalities and receptive fields of sensory neurones in the CNS of the leech. *Journal of Neurophysiology*, **31**, 740–56.

Nijenhuis, E. D. & Dresden, D. (1955). On the topographical anatomy of the nervous system of the mesothoracic leg of the american cockroach (*Periplaneta americana*). *Proceedings, Koniklijke Nederlandse Akademie van Wetenschappen*, **58**, 121–36.

Nordlander, R. H. & Singer, M. (1972). Electron microscopy of severed motor fibres in the crayfish. *Zeitschrift für Zellforschung und Mikroskopische Anatomie*, **126**, 157–81.

Nuesch, H. (1968). The role of the nervous system in insect morphogenesis and regeneration. *Annual Review of Entomology*, **13**, 27–44.

Pearson, K. G. (1972). Central programming and reflex control of walking in the cockroach. *Journal of Experimental Biology*, **56**, 173–93.

Pearson, K. G. & Bergman, S. J. (1969). Common inhibitory motoneurones in insects. *Journal of Experimental Biology*, **50**, 445–73.

Pearson, K. G. & Bradley, A. B. (1972). Specific regeneration of excitatory motoneurons to leg muscles in the cockroach. *Brain Research*, **47**, 492–6.

Pearson, K. G. & Iles, J. F. (1970). Discharge patterns of coxal levator and depressor motoneurons of the cockroach, *Periplaneta americana*. *Journal of Experimental Biology*, **52**, 139–65.

(1971). Innervation of the coxal depressor muscles in the cockroach, *Periplaneta americana*. *Journal of Experimental Biology*, **54**, 215–32.

Pearson, K. G., Stein, R. B. & Malhotra, S. K. (1970). Properties of action potentials from insect motor nerve fibres. *Journal of Experimental Biology*, **53**, 299–316.

Pipa, R. L. & Cook, E. F. (1959). Studies on the hexapod nervous system. I. The peripheral distribution of the thoracic nerves of the adult cockroach *Periplaneta americana*. *Annals of the Entomological Society of America*, **52**, 695–710.

Pitman, R. M., Tweedle, C. D. & Cohen, M. J. (1972). Branching of central neurons: intracellular cobalt injection for light and electron microscopy. *Science*, **176**, 412–14.

Potts, M. (1966). A rapid technique for graphic reconstruction. *Acta anatomica*, **65**, 315–21.

Pringle, J. W. S. (1939). The motor mechanism of the insect leg. *Journal of Experimental Biology*, **16**, 220–31.

Pusey, H. K. (1939). Methods of reconstruction from microscopic sections. *Journal of the Royal Microscopical Society*, **59**, 232–44.

Sahota, T. S. & Edwards, J. S. (1969). Development of grafted supernumerary legs in the house cricket, *Acheta domesticus*. *Journal of Insect Physiology*, **15**, 1367–73.

Sandeman, D. C. (1971). The excitation and electrical coupling of four identified motoneurons in the brain of the Australian mud crab, *Scylla serrata*. *Zeitschrift für vergleichende Physiologie*, **72**, 111–30.

Stretton, A. O. W. & Kravitz, E. A. (1968). Neuronal geometry: determination with a technique of intracellular dye injection. *Science*, **162**, 132–4.

Stuart, A. E. (1970). Physiological and morphological properties of motoneurones in the central nervous system of the leech. *Journal of Physiology*, **209**, 627–46.

Teutsch-Felber, D. (1970). Experimentelle und histologische Untersuchungen an der Thoraxmuskalatur von *Periplaneta americana* L. *Revue Suisse de Zoologie*, **77**, 481–523.

Usherwood, P. N. R. (1963*a*). Response of insect muscle to denervation. I. Resting potential changes. *Journal of Insect Physiology*, **9**, 247–55.

(1963*b*). Response of insect muscle to denervation. II. Changes in neuromuscular transmission. *Journal of Insect Physiology*, **9**, 811–25.

(1968). A critical study of the evidence for peripheral inhibitory axons in insects. *Journal of Experimental Biology*, **49**, 201–22.

Wigglesworth, V. B. (1959). The histology of the nervous system of an insect, *Rhodnius prolixus* (Hemiptera). II. The central ganglia. *Quarterly Journal of Microscopical Science*, **100**, 299–313.

Young, D. (1969). The motor neurons of the mesothoracic ganglion of *Periplaneta americana. Journal of Insect Physiology*, **15**, 1175–9.

(1970). The structure and function of a connective chordotonal organ in the cockroach leg. *Philosophical Transactions of the Royal Society of London*, B, **256**, 401–26.

(1972). Specific re-innervation of limbs transplanted between segments in the cockroach, *Periplaneta americana. Journal of Experimental Biology*, **57**, 305–16.

Young, D., Ashhurst, D. E. & Cohen, M. J. (1970). The injury response of the neurones of *Periplaneta americana. Tissue and Cell*, **2**, 387–98.

Zawarzin, A. (1924). Zur Morphologie der Nervenzentren. Das Bauchmark der Insekten. Ein Beitrag zur vergleichenden Histologie. (Histologische Studien über Insekten. VI). *Zeitschrift für Wissenschaftliche Zoologie*, **122**, 323–424.

The curious nature of degeneration and regeneration in motor neurons and central connectives of the crayfish

R. R. Hoy

Introduction

It is a widely accepted dogma in neurophysiology that the nucleated portion of the neuron, the cell body or soma, is the 'trophic centre' of the cell. This extension of the cell doctrine to nerve cells implies that all portions of a neuron, and in particular its lengthy axonal and dendritic extensions, are dependent upon contributions from its nucleated soma for maintenance. Isolation from the nucleus results in atrophy and death of the axon. Recent experiments in crayfish and some insects, that demonstrate great longevity of soma-less axons, suggest that we must modify our notions that neuronal somata act as biosynthetic centres from which metabolites are *constantly* being drawn, and shuttled down axons. This article will contrast the effects of axon lesions, primarily in motor neurons, between arthropods and vertebrates, which have been much more extensively studied. First it is appropriate to review briefly the evidence for the trophic centre concept from investigations in vertebrate nerves. The amount of experimental and clinical support for this concept is overwhelming and can be summarised under three points.

(1) Isolation of an axon from its cell body by cutting or crushing results in swift functional and morphological disintegration of the anucleate distal axon stump. In the vertebrate literature this is known as Wallerian degeneration, and has been acknowledged for over a century in experimental and clinical literature.

(2) Regeneration of a sectioned axon takes place by proximo-distal outgrowth of new nerve sprouts from the nucleated portion of the injured neuron. Ultimately outgrowth results in the replacement of the degenerated distal axon by a new distal apparatus, including synaptic terminals, i.e. re-innervation occurs.

(3) Proximo-distal transport of axoplasmic materials occurs and provides a mechanism by which substances synthesised in the cell

body can be conveyed to remote portions of the neuron such as the axon(s). This was first demonstrated by Weiss & Hiscoe (1948), and has been reviewed recently (Grafstein, 1969).

Although important aspects of the functional and anatomical organisation of neurons differ between vertebrate and invertebrate nervous systems, morphological and presumably physiological features of nerve cells themselves are quite similar. It was thus reasonable to expect cell bodies of invertebrate neurons to act as trophic centres for their distal axons. Injury-induced Wallerian degeneration does seem to occur in some arthropod axons, notably in the cockroach (Bodenstein, 1957; Hess, 1960; Guthrie, 1962, 1967; Jacklet & Cohen, 1967). However in other arthropods, particularly in crayfish, examples of slow neuronal degeneration have been reported (Johnson, 1926; Wiersma, 1960) but not systematically investigated. Recent work in locusts (Usherwood, 1963*a*, *b*; Boulton, 1969; Boulton & Rowell, 1969) demonstrates that sectioned axons remain viable for relatively long periods of time. Such studies and our own in crayfish (Hoy, Bittner & Kennedy, 1967; Hoy, 1968, 1969; Bruner, LeDouarin & Hoy, 1971) force us to question the general validity of the trophic centre concept.

Effect of axonal section on vertebrate and crayfish motor nerves

Most of what is known about neural degeneration and regeneration comes from studies of vertebrate nerves. In turn this work has served as a paradigm for parallel studies in invertebrate systems, so again we shall begin with a brief outline of the effects of axon section in vertebrates.

Vertebrates. If a motor nerve axon is interrupted by trans-section or crushing, degenerative changes occur both proximally and distally. By far the more severe changes occur in the distal, anucleate segment of the axon, although retrograde effects also occur and have been exploited as a means of tracing neural pathways. Usually synaptic transmission ceases within 24 h; the site of failure in frog neuromuscular junctions is the synapse (Titeca, 1935) and the precise time course of degeneration is temperature sensitive (Birks, Katz & Miledi, 1960). A few days later the axon begins to disintegrate, resulting in conduction failure and it is ultimately reduced to a hollow shell, usually within a few weeks. There is a concomitant degeneration of the myelin sheath. These events are typical of Wallerian degeneration.

Regeneration is accomplished by re-innervation, the proximo-distal outgrowth of nerve filaments from nucleated cell bodies. Ultimately this results in the synthesis of an entirely new set of distal axons and neuromuscular junctions.

Crayfish. Crayfish motor nerves do not undergo classical Wallerian degeneration when injured. The distal portion of sectioned axons do not degenerate for many months following lesion. Physiological and ultrastructural investigations demonstrate that isolated axon stumps persist in apparently normal condition for more than a half-year after the initial lesion was made, in the absence of regeneration. Regeneration, when it occurs, is rapid, often within one month after lesion. Physiological evidence suggests that regeneration involves the fusion of proximal and distal axon segments rather than re-innervation by proximo-distal outgrowth and the establishment of an entirely new set of neuromuscular junctions. These conclusions are drawn from previously published studies (Hoy *et al.* 1967; Hoy, 1968, 1969; Bruner *et al.* 1971) and unpublished studies (Bruner & Hoy, in preparation) that will be summarised below.

Neuromuscular systems

Absence of Wallerian degeneration in crayfish motor nerves

Claw preparation. A visit to an aquarium will demonstrate that disturbing or threatening a lobster or crayfish results in a defensive raising and opening of its claws. The claw opener muscles are innervated by a single excitatory neuron whose cell body lies in the first thoracic ganglion and which sends its axon along the length of the claw. When the nerve trunk containing this axon is severed or crushed, the animal is no longer able to raise and open its claw when irritated with a brush about its eyes or mouthparts; for our purposes it was important that defensive behaviour could normally be elicited via afferent reflex pathways independent of the claw itself. We could thus measure the time required for regeneration by simply noting when recovery of the opening reflex occurred. It was also possible to test the functional state of the isolated distal

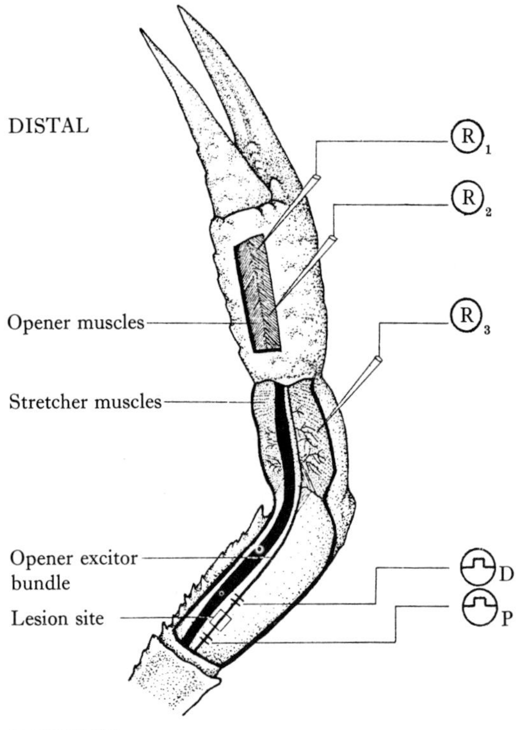

Fig. 1. Crayfish claw prepared for electrophysiological tests of regeneration and/or degeneration. Microelectrodes, R, are inserted along the innervation pathway of the opener excitor nerve bundle. The square wave symbols refer to stimulation sites distal, D, and proximal, P, to the site of the nerve lesion. Note that the opener excitor nerve innervates two sets of muscles, the stretcher (of the carpopodite) and more distally, the opener muscles, and that the innervation path is linear (see text). (From Hoy *et al.* 1967. Copyright 1967 by the American Assoc. Advancement of Science).

measured by applying behavioural and physiological tests. The stump of the opener excitor nerve bundle in non-regenerated claws. This was done by stimulating it electrically while recording intracellularly from single fibres of the opener muscle (Fig. 1). In the normal intact claw this produced depolarising excitatory junctional potentials (EJPs), one for each stimulus pulse applied to the nerve (Fig. 2*a*). The opener excitor nerves were cut in a large number of animals and progress toward regeneration was

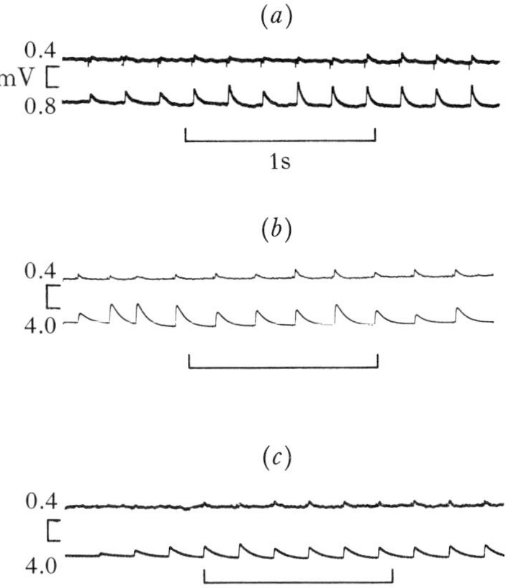

Fig. 2. Junctional potentials recorded intracellularly from muscle fibres of the claw opener in response to stimulation of the opener excitor axon at 5 stimuli/sec. (*a*) Control nerve, not operated. (*b*) Regenerated nerve (after nerve crush) 31 days post-operative. Nerve stimulated proximal to the site of lesion. (*c*) Non-regenerated, soma-less nerve 51 days after the opener excitor nerve bundle was sectioned. The nerve was stimulated distal to the lesion (the disconnected proximal stump was stimulated but had no effect on the opener fibres). In each example, the topmost trace is a recording from the distal-most fibres of the opener muscle, which produce small, non-facilitating EJPs at low stimulus frequencies. The lower trace is from proximal fibres of the opener muscle, which are characterised by large, facilitating EJPs at low frequencies. The differential response of muscle fibre junctional potentials with stimulation frequency is explained in the paper by Bittner (1968). (From Hoy *et al.* 1967. Copyright 1967 by the American Assoc. Advancement of Science.)

most interesting results came from animals in which neither degeneration or regeneration had occurred. Degeneration was not detected until after 100 post-operative days and out of three claws tested 140–160 days post-operatively, only one had degenerated. Degeneration was defined *physiologically*: the inability to record EJPs from muscle fibres during electrical stimulation of the distal nerve stump. Recently Bittner (personal communcation) has extended the functional survival of isolated opener nerves to over 200 days. Until degeneration finally occurred, synaptic properties were judged to be completely normal by such criteria as the amplitude of resting potential, presence of spontaneous miniature EJPs, and amplitude, waveform, and facilitation properties of electrically evoked EJPs (Fig. 2*c*). Experimental stimulation of these nerves often ran for hours without any sign that a small 'residual' supply of transmitter had been exhausted. In very old preparations 'poised' to degenerate, the initially excitable nerve evoked EJPs for a short time and then only intermittently. The site of failure was apparently the axon, which no longer supported spike conduction (Bittner, personal communication).

Abdominal flexor preparation. Out of water a startled crayfish will raise its claws and open them as a defensive gesture. In water it is as likely to react by swimming away, in a series of rapid 'tail-flips'. Swimming behaviour is mediated by the deep flexor muscles of the abdomen. These segmental muscles are innervated by a set of 10 motor neurons (in each half segment) which comprise the whole of the deep branch of the third root. The junctional physiology of this system in the third abdominal segment has been investigated earlier (Kennedy & Takeda, 1965*a*, *b*; Bruner & Kennedy, 1970). We examined the effect of cutting or crushing the third root upon the junctional properties in this muscle (Bruner *et al.* 1971). Our findings are similar to those in the claw. Degeneration resulting from nerve section, and even ganglionectomy, did not occur before 120 post-operative days; before this time the electrophysiological properties of neuromuscular junctions were normal as judged by the amplitude of resting potential, and the amplitude and waveform of electrically evoked EJPs. These junctions are especially interesting because of their complex synaptic properties following repetitive stimulation, producing various after-effects that affect synaptic efficacy of subsequent stimulation (Bruner & Kennedy, 1970). The effect of nerve section upon these junctions was investigated because we felt that the absence or persistence of such synaptic after-effects would indeed reveal how 'normally' axons function when isolated from their central connections for many weeks. In particular, one of the excitatory motor neurons, the 'motor giant' (MG), differs from the other excitatory cells in the root ('non-giants') in its response to repetitive stimulation. Normally, stimulation of the MG neuron by a train of stimuli delivered at a slow rate of 1 min^{-1} results in marked antifacilitation of EJPs, i.e. successive decrement of EJP amplitude until only a small amplitude EJP is produced by each stimulus pulse. This property is retained in a normal state even after 100 post-operative days. The other excitors (non-giants) also show no change in EJP amplitude to such slow frequency stimulus trains in either normal or 100-day post-operative animals. If the rate of stimulation is increased to 1–5 s^{-1} non-giant motor junctions respond with noticeable facilitation (augmentation of EJP amplitude with succeeding pulses of stimulation until a maximum amplitude is attained). Non-giant junctions respond this way in both intact controls and 100-day preparations. Thus whatever the underlying cause of these synaptic after-effects, it is not quickly affected by axonal section. Degeneration occurred gradually rather than abruptly in this system. Preparations tested around 120 post-operative days showed signs of deterioration in the form of greatly diminished EJP amplitudes (down to 1 mV, greater than a 10-fold reduction from normal), however such diminished EJPs were recorded in many muscle fibres in these older preparations. It was our impression that degeneration in the claw was not marked by the onset of reduced EJP amplitude; however it should be noted that a normal size EJP in claw muscles is only about 1 mV, much smaller than in the flexor.

Ultrastructure of severed motor nerves in crayfish

The information reviewed so far comes primarily from physiological studies. Recently, Nordlander & Singer (1972) applied electron microscopy to sectioned motor neurons in crayfish. Their results are particularly relevant because they selected the same nerves in the crayfish claw as those studied physiologically, the nerve bundle of the claw opener excitor.

They confirmed the earlier finding that degeneration is greatly retarded in sectioned motor axons; out of over 300 axons examined at various post-operative stages, only four actually degenerated (which occurred in the first four weeks). All others were normal even after 100 days, as judged by the appearance of the endoplasmic reticulum, longitudinally oriented microtubules, and oriented, elongate mitochondria. An especially interesting observation is that glial cells associated with severed axons undergo a striking change in morphology. The glial sheath doubles its thickness within 48 h, a condition which persists for months. The most marked changes occurred in the glial layer immediately adjacent to the axon, the adaxonal layer, where they observed greater accumulations of mitochondria, endoplasmic reticulum, and ribosomes than in other layers.

Regeneration in crayfish motor nerves

The test for regeneration was recovery of function. In the claw a simple behavioural test was applied to check for recovery of the claw opener; in the abdominal flexor muscles no behavioural tests for recovery were known, and instead electrophysiological tests were applied at regular intervals to individuals taken from a large group of nerve-lesioned animals. The nerve was exposed by dissection so that stimulating electrodes could be applied to the nerve *proximal* to the original lesion and the effect of stimulation measured with microelectrodes in distal muscle fibres. We could thus test for the regeneration of axons past the lesion which had made functional peripheral connections. This physiological test was also applied to regenerated claws.

The main features of regeneration are: (1) it often occurred very rapidly, sometimes within two weeks, and long before degeneration was expected to occur in the distal stump, and (2) electrophysiological properties of neuromuscular junctions in regenerated nerves were completely normal.

Regeneration of the opener excitor nerve of the claw. Regeneration occurred from 10 to 20 days following nerve crush and from 20 to 30 days following nerve section. As soon as an animal demonstrated recovery of its opener, its claw was removed and tested physiologically. Fig. 2*b* shows that EJPs due to electrically evoked stimulation are not different from normal ones. In all cases of regeneration, the two previously separated stumps appeared to be reconnected; no collateral nerve trunks were observed to have innervated the muscles of the claw dactyl. Recovery of function in the opener was always accompanied by extension of the carpopodite, this important observation will be discussed in relation to the mechanism of regeneration.

Regeneration of motor neurons of the abdominal flexor muscles. The time course of regeneration in this system was similar to that in the claw, i.e. after about 20–30 post-operative days. The criteria for regeneration was based entirely upon electrophysiological tests: the ability to stimulate a reconnected nerve trunk 'upstream' from the site of lesion and to record evoked EJPs from muscle fibres 'downstream'. The results were like those from the claw in that no differences could be discerned between regenerated and normal preparations on the basis of junctional properties. Visual examination of regenerated preparations were made and we did not observe innervation of muscles by nerve trunks other than the original one, except in three cases which will be discussed separately below. Whatever the mechanism of regeneration, the original distal stump was utilised in regeneration.

Ultrastructure of regenerating nerves (Nordlander & Singer, 1972). The time course of regeneration in their preparations are consistent with our own. They used the same behavioural test for recovery of the opener reflex and also confirmed that recovery occurs more rapidly from crushed nerves than cut ones. Beginning after four post-operative weeks they found 'double-membrane-bound compartments' within the glial sheath which appeared to be axonal in ultrastructural morphology; they contained oriented microtubules and vesicles that resembled synaptosomes. These compartments were termed 'satellite axons'. These satellite axons appeared to be outgrowth filaments from the proximal nerve stump. Not only did they resemble axons in morphology but they appeared first in the more proximal portions of the distal segment and apparently grew distally into the segment, within the glial sheath. These filaments were apparently not followed in the distal stump beyond 1.5 cm or as far as the muscles.

The mechanism of regeneration: axonal fusion

The distal stump of a sectioned crayfish axon can remain viable for many months without apparent functional deficit, yet whenever distal and proximal nerve stumps meet regeneration occurs within weeks. It is thus possible that regeneration might simply involve a reconnection of proximal and distal axons by fusion. Whether the axoplasm of the two stumps actually becomes confluent or whether a membrane persists at the site of fusion cannot be distinguished by our electrophysiological experiments. Whatever the case it seems unlikely that regeneration occurs by the universally accepted mechanism of proximo-distal outgrowth from the proximal stump that results in the establishment of an entirely new set of synaptic terminals at the muscle. While overlapping time courses of regeneration and prolonged survival of the distal axon permit us to entertain such a novel mechanism of regeneration as fusion, it does not prove it. We will now review the physiological evidence that supports the fusion hypothesis (Hoy *et al.* 1967).

Evidence from studies in the claw. Two lines of physiological evidence support regeneration by fusion; both are based on experiments which have two outcomes, one of which supports fusion whereas the other supports the conventional outgrowth mechanism.

1. Electrical stimulation of both distal and proximal portions of the regenerated axon produce only a single EJP at a sharply defined stimulation threshold. The claw opener is innervated normally by a single excitatory axon which produces a single unitary EJP for each axon spike. In newly regenerated claws stimulation of the axon *proximal* to the lesion should have produced only a single EJP (indicating functional presence of a single excitor) whether regeneration had occurred by fusion or outgrowth. However, stimulation of the *distal* stump should have produced only a single EJP if regeneration had occurred by fusion, but *two* EJP types if regeneration had occurred by axonal outgrowth, because regeneration occurs within one month and the distal stump is capable of surviving at least 3 months or more. Thus, outgrowth should *add* an additional excitor in the distal stump; its presence should have been detected by a graded series of stimuli applied to the axon, which would have been activated independently of the original nondegenerated axon, and the newly added outgrowth axon. In fact, one and only one JP was ever produced by stimulation of the distal stump; no evidence for dual excitatory innervation was ever found in our experiments on the claw.

2. Regeneration resulted in abrupt recovery of function at all points along the path of innervation of the opener excitor nerve. If regeneration were occurring by proximo-distal outgrowth, recovery of function should be gradual and correlated with the slow outgrowth of the nerve along the innervation pathway, during which new terminals are being laid down, i.e. first proximally, later distally. Fusion on the other hand would result in a sudden recovery of function at all junctions because it only requires re-uniting the still functional distal apparatus with the proximal portion of the neuron which retain central reflex connections. It

was possible to test these opposing mechanisms in the claw because of a fortuitous bifunctional innervation pattern of the opener nerve, and the fact that the innervation pathway is linearly arranged in a proximo-distal orientation (Fig. 1). The opener excitor nerve of the dactyl segment is also the excitor for the stretcher muscles of the carpopodite segment of the claw; this segment is proximal to the dactyl. Thus activation of this axon results in extension of the carpopodite as well as opening of the dactyl. Depending on the size of the claw the distance from the proximal-most stretcher muscle fibres to the distal-most fibres could be up to 40 mm. Fig. 1 shows that the innervation field of the opener excitor neuron is linear. Thus, if regeneration occurred by outgrowth one would expect recovery of the stretcher reflex *before* the opener reflex. In fact, recovery of function of the stretchers always accompanied, but never preceded claw opening. These behavioural observations were supported by intracellular recording from stretcher and opener fibres simultaneously, while stimulating the newly regenerated axon proximal to the lesion site (Fig. 1). A sampling of numerous fibres from both stretcher and opener muscles in each of eight preparations which had just regenerated showed that EJPs appeared simultaneously in both muscles at the same threshold value of nerve stimulation, implying that only one axon was activated that innervates both muscles.

Evidence from regeneration in the abdominal flexor system. The data have been presented elsewhere (Hoy, 1969) and will be summarised below. The superficial flexor system is simple (50–75 muscle fibres are innervated by only six motor neurons) and functionally mapped so that identification can be made between activity of the six identifiable motor neurons and the type of EJPs their activity produces (Kennedy & Takeda, 1965*a*, *b*). It is a tonically active system; ongoing spontaneous junctional activity is the normal state in this muscle. A preparation in which the nerve containing the six excitors had been interrupted 45 days earlier was tested for regeneration. Only one of the six motor units had regenerated (Fig. 3*a*) and evidence of tonic or reflexively activated activity from the other five units was not found. (The single ascending motor neuron was crushed at the time of the experiment.) However, electrical stimulation of the root evoked activity from previously 'silent' units demonstrating that they had not

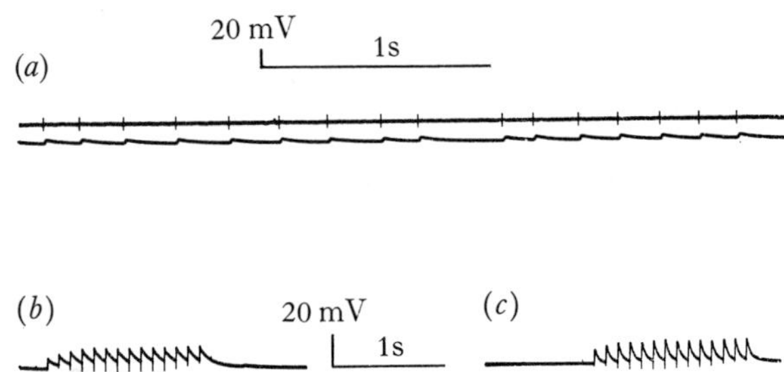

Fig. 3. Recordings from superficial third root and slow flexor muscle fibres 45 days after the motor axons were crushed in the III–IV connective. (*a*) simultaneous recording of spontaneous discharges in the regenerated motor axon (top trace) and junctional potentials (bottom trace), evoked by activity of a lone regenerated motor axon, in a single fibre of the slow flexor muscle. (*b*), (*c*) intracellular recordings from two different muscle fibres showing JPs evoked by *electrical* stimulation of the third root, but not evoked by spontaneous discharge in the nerve. The activity in (*a*) cannot be attributed to the single ascending motor neuron from ganglion IV, which was crushed at the time of the recording. Calibrations: (*a*), 1 s and 20 mV; (*b*), (*c*), 1 s and 20 mV; voltage calibrations refer to recordings of EJPs only. (From Hoy, 1969.)

degenerated, although regeneration clearly had not occurred (Fig. 3*b*, *c*). Thus, the root contained axons whose peripheral connections were intact, but not their proximal connections. This situation would not have been expected if regeneration had occurred by re-innervation. It is perfectly compatible with regeneration by fusion. The fact that only one neuron had regenerated is not surprising because fusion would not be expected to have occurred simultaneously for all six axons. Visual examination of the methylene blue stained preparation revealed that the root contained only the normal number of axons – six.

Hyper-innervation in the deep flexor system

These experiments are the result of collaboration with Dr Jan Bruner and have not been reported elsewhere. A number of cray-fish were operated upon in such a way that the third roots of the connective between the third and fourth abdominal ganglia were completely isolated from both ganglia (Fig. 4*b*). The axons of the deep flexor motor neurons in this resected connective were thus completely isolated from their cell bodies, most of which reside in the third ganglion and the remainder in the fourth ganglion. In

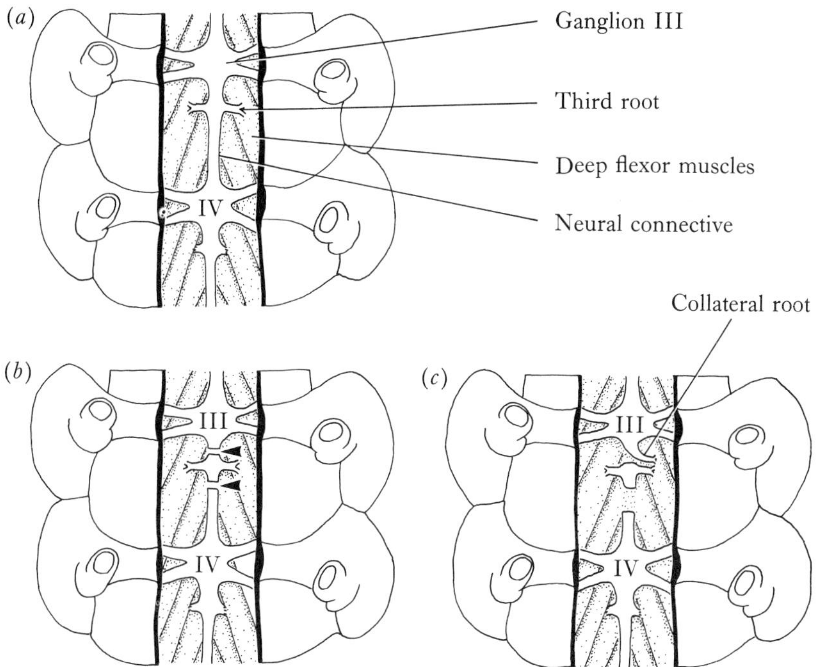

Fig. 4. Sequence of events leading to hyperinnervation of the deep flexor muscles in the crayfish abdomen. (*a*) the situation before the operation, with the third root emerging from the III–IV connective midway between the ganglia. (*b*) the results of the operation; a section of the connective containing the axons of the third root is completely isolated, by means of two cuts (arrows), from both ganglia (which contain the cell bodies for the third root axons). (*c*) the situation two months after (*b*). The original third roots are still connected to the flexor muscles, but in addition, the muscles are innervated by a new collateral root, that is connected to the third ganglion. These sketches are 'cut-aways' for illustrative purposes. In the actual operation (*b*), the connectives were not exposed to such a great extent. The swimmerets were amputated for clarity of illustration.

three such animals all between 50 and 70 days post-operative, the deep flexor muscles of the third abdominal segment were found to be innervated, not only by the still intact third roots of the resected connective, but also by a new collateral 'root' which had grown out proximally from the connective stump of the third ganglion (Fig. 4*c*). These fortuitous preparations enabled us to answer several questions regarding regeneration. (1) Can regeneration by re-innervation occur in crayfish? (We presented evidence for fusion but we do not *exclude* regeneration by outgrowth.) (2) Can the original third root (which normally persists for over 100 post-operative days) retain its functional capacity in the presence of new terminals from axons that are connected to cell bodies in the ganglion? (3) Are given, individual neuromuscular junctions on the flexor muscle fibres shared between the new outgrowth axons and the original axons, or are they functionally independent?

These questions were answered by applying electrophysiological tests. Junctional potentials due to electrical activity in either root was detected by intracellular recording from single flexor muscle fibres whilst the original third root and/or the new collateral root was electrically stimulated. By stimulating each root separately and recording the effect upon single muscle fibres we observed that both roots contain axons that are functionally connected to the flexor muscle. Fig. 5*e*, *f* shows that EJPs can be recorded from the same muscle fibre as a result of independent stimulation of both roots. Many muscle fibres were 'dually innervated'. Thus it is clear that the presence of an old (but functionally persistent) set of axons does not prevent a new set from growing in from the central nervous system and innervating the muscle. Conversely, the establishment of synaptic connections by new axons that are connected centrally to cell bodies does not disrupt the function of junctions of the original axon, lacking in nuclear connections; new outgrowth axons certainly do not 'kill' the terminals of the original axons, nor render them non-functional.

As for whether or not the neuromuscular junctions of old and new roots are different from each other, a physiological test for independence was applied as follows. Single, suprathreshold stimulus pulses were applied to both roots separately with a slight delay between them; junctional activity was monitored by a microelectrode in single muscle fibres. Junctional potentials in the same muscle fibre were evoked from separate stimulation of the new collateral and the original nerve root. The results of interacting stimulus pulses from the two roots are shown in Fig. 5*a–d*. Initially the stimuli were applied about 250 ms apart and each pulse produced a single, independent EJP. The electrical properties of the two junctions are distinctly different, as can be seen in their response to repetitive stimulus shocks. Fig. 5*e*, *f* shows that the EJPs from the original third root show antifacilitation to repe-

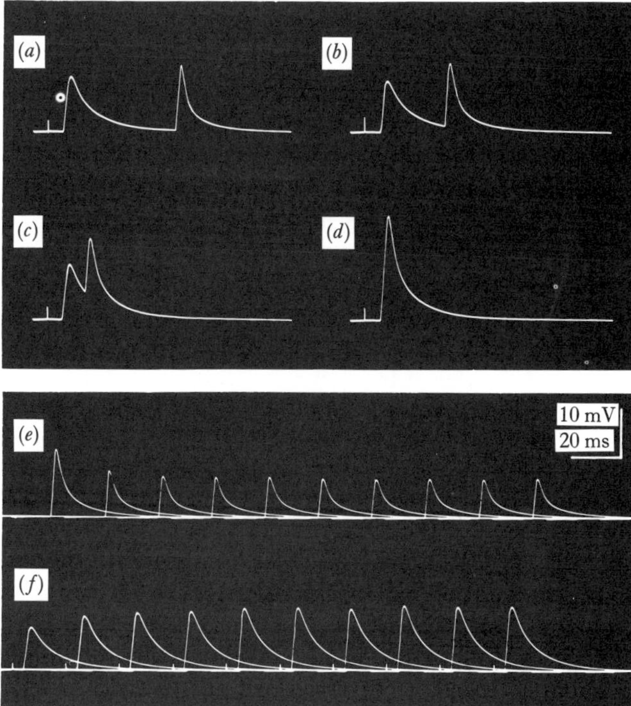

Fig. 5. Junctional potentials from a hyperinnervated, single muscle fibre of the deep flexor muscles. (*a–f*) recordings from the same muscle fibre during different combinations of stimulation of the original third root and the new, collateral root, both of which innervate the fibre. (*a–d*) temporal interaction of EJPs evoked by electrical stimulation of the collateral and original roots culminating in summation, in (*d*), due to simultaneous stimulation of both roots. (*e*) stimulation of the original third root by itself, (*f*) stimulation of the new collateral root by itself.

titive stimulation whereas the EJPs of the new root facilitate. As the delay between stimulus pulses was decreased the EJPs began to fuse (Fig. 5*c*), so that the EJPs were superimposed upon each other. When both nerve roots were stimulated simultaneously a compound EJP was evoked whose amplitude was greater than either unitary EJP (Fig. 5*d*). Strict arithmetic summation did not occur, presumably because the amplitude of the compound EJP was limited by the equilibrium potential for the ion species active in depolarisation. If the two nerve roots shared common neuromuscular junctions, their simultaneous stimulation should not have resulted in a summated potential. It thus seems likely that the two roots were not only anatomically separate but also have independent synaptic terminals. Morphological confirmation might have been provided by electron microscopy but this was not possible at the time. These experiments and others (manuscripts in preparation) demonstrate that regeneration by classical proximo-distal re-innervation can occur in crayfish motor systems, and that the junctional properties of such nerves appear to be quite normal. However, the presence of innervation from nerves connected to cell bodies does not affect the persistence of functional innervation from the original axons which have been long isolated from cell bodies.

It would be of interest to know the identity of the axons in each of the roots which dually innervate a given flexor muscle fibre. Since only a few of the axons in the third root actually innervate any given muscle fibre it is possible that such fibres are innervated by axons in the original root and the new outgrowth root which share common cell bodies in origin. This would suggest that persistent axons in the original root prevent the formation of synapses by 'foreign' motor neurons, but allow synaptogenesis by homologous neurons. Such questions are of interest in regard to the 'trophic' function of nerves. In the crayfish the persistence of synaptic activity in the absence of a cell nucleus might or might not be related to persistence of trophic functions which might depend more upon somatic contributions. Unfortunately our data were not sufficient to allow specific identification of motor neurons but such experiments are within the realm of experimental possibility. These remarks are speculative, however, since almost no work has been done in regard to trophic functions in crayfish neuromuscular systems.

The central connectives

Regeneration in the abdominal connective

All of the experiments described above were conducted on neuromuscular systems, i.e. the axons lesioned were motor neurons. A few experiments were also performed on the axons of the central connectives to examine regeneration of sensory and/or interneurons (Hoy, 1968).

Axons that course through the connective between the fifth and sixth abdominal ganglia were interrupted as shown in Fig. 6;

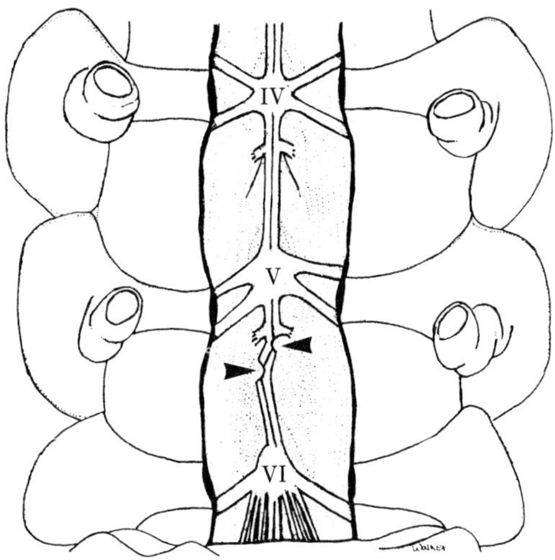

Fig. 6. Diagram of the lesion made in the V–VI connective which interrupts all axons in the connective. To insure that *all* axons are interrupted, the cord was sectioned three-quarters through at two levels on opposite sides in the connective (arrows). Swimmerets were amputated for illustrative purposes.

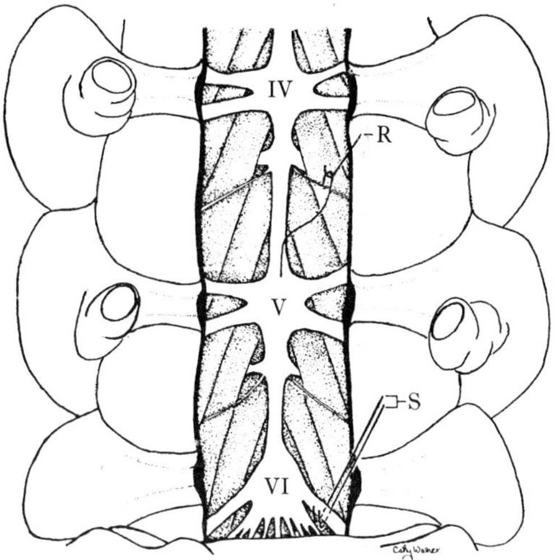

Fig. 7. Experimental procedure to test for regeneration in connectives after operation shown in Fig. 6. Stimulating electrode, S, placed on roots of the sixth ganglions. Fibre bundles are stripped from the IV–V connective and placed on recording electrodes, R.

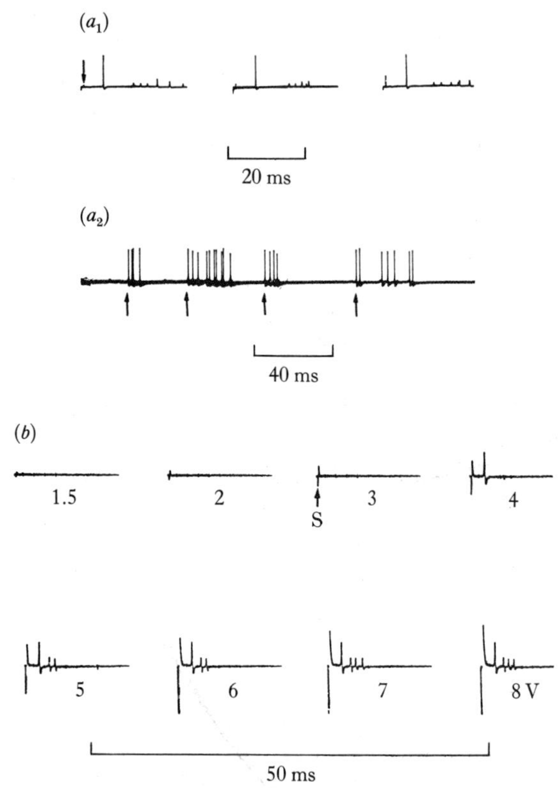

Fig. 8. Electrophysiological data produced from the procedures of Fig. 7. In (a_1), electrical stimulation of a sixth ganglion root results in one-to-one driving of a unit isolated in a nerve bundle in the IV–V connective (arrow, stimulus artifact). This same unit can be driven synaptically by stimulation of cuticular hairs of the tail (a_2; arrows indicate application of tactile stimulation to tail). In (b), a different set of units are isolated in another bundle in the IV–V connective. The numbers refer to voltage steps applied to sixth ganglion roots required to recruit regenerated units in the IV–V bundle. Note that the units are recruited at a definite threshold, at constant latency, and one-to-one with single stimulus pulses. Arrows refer to stimulus artifact associated with single pulses of electrical stimulation to ganglion VI roots.

several dozen such operations were made at one time. Later, individuals were sampled at periodic intervals for physiological tests of regeneration. The experimental procedure is shown in Fig. 7. Stimulating electrodes were applied to the roots of the sixth ganglion, and axonal filaments (containing several axons) were stripped away in the IV–V connective and placed upon recording electrodes. The criterion for regeneration was the presence of units which responded in a 'stimulus-locked' manner, i.e. recruited by a stimulus pulse at a sharply defined threshold and following 1:1 with each stimulus pulse. Whenever possible the tactile receptive fields of active units were mapped as well. In normal control preparations these tests identify the presence of primary sensory units and interneurons. In our regeneration tests we were concerned with any unit which could be driven 1:1 by electrical stimulation of the roots of the sixth ganglion.

The time course of regeneration is similar to that in neuromuscular systems. Between 18 and 30 days post-operative time were required; the older the post-operative age of the preparation the greater the number of active units recorded. In the earliest preparation found to regenerate (17 days), only three units were active. In older preparations, nearly every filament isolated contained several active units. The actual number of active units in each regenerated preparation was not tabulated; an estimate would be around six units in early regenerated connectives to dozens in older ones. A typical result is shown in Fig. 8*b* in which a series of increasingly stronger stimulus pulses are applied to the roots of the sixth ganglion. Several stimulus-locked units are seen to be driven by stimulation. Fig. 8*a* shows a regenerated unit that responds to single pulses of electrical stimulation (a_1) and to tactile stimulation applied to cuticular hairs of the telson, by giving short bursts of spikes whenever hairs were touched (a_2, arrows).

We are not able to say much more than that regeneration of axons in the central connectives unquestionably occurs, and that it requires at least 20 to 30 days. An interesting question remains as to the *specificity* of regeneration in these axons.

Degeneration in the central connectives

The recent work of Wine (personal communication) is especially illuminating because he obtained a continuous measure of the course of degeneration following nerve section. The easily identifiable medial giant interneurons of the crayfish nerve cord were studied. These paired neurons run the length of the entire animal and are among the largest in the nervous system (over 100 μm in diameter); each is a single cell whose soma lies in the supra-oesophageal ganglion. After they were sectioned, distal axons remained functionally potent, as measured by the ability to propagate impulses and transmit synaptically, for nearly six months. In this respect they behave like the distal axon of sectioned motor neurons. Degeneration eventually occurs, although gradually, in the form of diminished conduction velocity due to decreased cross-sectional area of the axon. Wine also observed profound retrograde effects in the proximal axon of the sectioned neuron, particularly a marked hypertrophy of the proximal segment. Regeneration was not observed to occur in these giant interneurons.

Discussion

Neural degeneration in insects and crayfish

The time required for degeneration to occur in the distal stumps of transected arthropod nerves is highly variable. The persistence of crayfish motor axons *sans soma* is of heroic proportions – over half a year (Table 1). In insects, survival times are more modest and are complicated by factors such as temperature and regeneration. Thus far two weeks to one month seems to be the limit of survival before either degeneration or regeneration intervenes. We shall compare the effects of nerve lesion on distal axons in insects and crayfish and then follow with a discussion of factors that might operate to maintain isolated crayfish axons.

The effects of axon section

In motor nerves the time required for degeneration to occur following axon section is variable, even between as closely related insects as cockroaches and locusts. Degeneration following lesion occurs in the cockroach with a rapidity and a course of events that are very similar to Wallerian degeneration in vertebrate nerves (Bodenstein, 1957; Guthrie, 1967; Jacklet & Cohen, 1967). Jacklet & Cohen provided electrophysiological data showing that both junctional transmission and the spontaneous occurrence of miniature EJPs cease within 3 to 5 days after lesion of motor neurons of the metathoracic walking legs. Transmissional breakdown was correlated with a decrease in the amplitude of the resting potential in denervated muscle fibres by as much as 30 per cent. Roeder & Weiant (1950) found that direct electrical stimulation of denervated leg muscles resulted in contractions for 3–5 days post-lesion, but after this time the muscle apparently becomes

Table 1. *Survival time of isolated crayfish axons*

Preparation	*Days*
Claw opener (neuromuscular)[1]	
Hoy *et al.* (1967)	> 140
Nordlander & Singer (1972)	150
Abdominal deep flexor (neuromuscular)[1]	
Hoy (1969)[2]	> 40–50
Bruner *et al.* (1971)	120
Abdominal superficial flexor (neuromuscular)[1]	
Hoy (1969)[2]	> 40–50
Central connectives (medial giant fibres)	
Johnson (1926)[2]	*ca.* 40
Wine (pers. comm.; in preparation)	150

[1] Survival is defined as persistence of synaptic transmission.
[2] Degeneration was not followed to conclusion; these are the limits are far as the experiments were carried out.

electrically inexcitable until regeneration occurs some weeks later (Bcdenstein, 1957). In the locust, however, Usherwood (1963*a*, *b*) found that the distal segments of lesioned motor nerves maintained their function considerably longer than in the cockroach. Neuromuscular transmission was 'maintained' for nearly a month in some cases. However, unlike the case in crayfish, abnormalities in the electrical characteristics of the junctional potentials appeared soon – between three and nine days – after lesion. In particular, electrically evoked EJPs became steadily diminished in amplitude and sometimes their waveform became uncharacteristically distorted. The spontaneous release of transmitter, measured by occurrence of miniature EJPs, persisted for a month but also became highly atypical. Usherwood correlated the loss of junctional transmission with marked decreases in the amplitude of the resting potential of the muscle fibres innervated by the injured nerve. This is in striking contrast to the crayfish, where no reductions in resting potential in muscle fibres occur as a result of axon section (Bruner & Hoy, unpublished results); on the contrary, resting potentials have been reported to increase as a result of motor nerve section (Girardier, Reuben & Grundfest, 1962). The situation in locusts if further complicated by the fact that Usherwood, Cochrane & Rees (1968) reported that the survival of the isolated distal nerve stump is dependent upon temperature. By elevating the ambient temperature to 30 °C neuromuscular synaptic transmission failed within two days; previous experiments in which transmission persisted for 9–24 days were performed at 20 °C. This is reminiscent of the results of Birks *et al.* (1960) who found that lesioned motor axons in frogs survived longer if the frogs were stored at low temperatures than at higher ones. Temperature effects are not to be doubted, but the relative effect on survival time must be considered. Holding animals at much lower temperatures may prolong axon survival for days, and even a few weeks, but crayfish axons survive for many months. Although the matter has not been studied systematically, I have held nerve-lesioned crayfish at various temperatures ranging from 12 to 27 °C without noticing abrupt changes in axon survival times.

Turning from motor nerves to axons in the central connectives, the result of transection again varies depending upon the species of insect studied. In the cockroach, transection results in rapid degeneration; these results are based on both physiological and ultrastructural investigations (Hess, 1960; Farley & Milburn, 1969). Hess found degeneration to occur so rapidly that axons seem to have disappeared within a week after section. On the other hand axons in the central connectives of locusts and mantids seem much more resistant to degeneration after nerve section. In a series of ultrastructural studies, Rowell & Dorey (1967), Boulton (1969), and Boulton & Rowell (1969) examined the effect of transection in the thoracic and neck connectives of *Schistocerca* and *Sphodromantis*. Only about 2 % of the axons showed degeneration and the majority of the axons showed 'reactive' changes (clearly not degenerative) that persisted until regeneration began to occur, about a week after lesion. An intermediate situation has been described in the moth, *Galleria mellonella* by Tung & Pipa (1971). A fraction of the axons in cut central connectives seem to persist for nine days without degenerative signs; however these seem to be exceptional, because most of the other axons degenerate progressively after nerve lesion. These results in insects stand in contrast to the results of Wine (personal communication, manuscript in preparation) who investigated the effect of axonal transection of the medial giant interneuron in the central connectives of the crayfish. The distal, soma-less, portion of the neuron required up to five months before degeneration finally occurred.

Before this time, the distal axon could still support impulse conduction and synaptic transmission in a relatively normal condition, although the conduction velocity of the axon spike decreased, a result which was correlated with diminishing cross-sectional area of the distal axon with time.

The reaction of glial cells in sectioned axons is also quite variable in insects. In the cockroach, Hess (1960) noted an increase in the glial sheath associated with degenerating axons; it was not determined whether the increase was due to proliferation, hypertrophy, or migration of glial cells. In the locust, Boulton (1969) found that glia seem to behave independently of the fate of their associated axons in the severed connective non-degenerating axons might be surrounded by degenerating glia, or degenerating axons might be associated with normal, non-degenerating glia. Again, it was not known if glia were proliferating *in situ* or migrating from adjacent areas. As will be discussed below, neuroglia in crayfish distal axons react to lesion with cellular hypertrophy.

How is degeneration being prevented or delayed in the central connectives of locusts, mantids, and crayfish, and the motor neurons of crayfish? Degeneration involves two events. First the axoplasm itself begins to disintegrate, and this is followed by the literal invasion of the axon by phagocytes which devour the axoplasmic debris. The actual mechanisms that trigger these events are not known. Let us first consider the situation in the locust, *Schistocerca gregaria* and the praying mantis, *Sphodromantis* sp., which were studied by Boulton & Rowell (1969). Neck connectives were cut midway between suboesophageal and prothoracic ganglia; this resulted in severing about half of the axons from their cell bodies. However only about 2 % of the axons actually degenerated for up to 23 post-operative days. Many of the non-degenerating axons show 'reactive' changes and about 2 % of these reactive axons were shown to contain osmiophilic granules 30–100 μm in diameter. Such granules were also found in newly regenerating axons. If, on the other hand, the neck connectives were cut in two places resulting in the isolation of *all* axons from their cell bodies, degeneration was swift for all axons in the isolated segment. Degenerative signs appear by the second day, and by the ninth day the isolated connective, now shrunken in size, is filled with phagocytes. No granules were seen in these degenerating axons. The authors suggest that these granules are formed in response to injury within axons that are still attached to their nucleus. (Since a totally isolated connective segment has no nuclear connections granules would be expected to be absent.) The role of these presumably neurosecretory granules might be in the active inhibition of degeneration and phagocytosis. Such granules have also been observed in cut motor nerves of the cockroach, *Periplaneta* (Milburn, 1971). These granules were *only* seen in regenerating (nucleated) neurons. This interesting hypothesis for the 'trophic' role of granules invites further study.

The results in locust and mantis can be compared with the crayfish. Nordlander & Singer (1972) in their ultrastructural study of the opener nerve bundle of the crayfish claw, were unable to find osmiophilic granules of the type described by Boulton & Rowell. The opener nerve bundle is a mixed nerve, consisting of motor axons and many sensory axons. In a sectioned nerve, therefore, many of the axons (sensory) would be still connected to their nuclei. However Nordlander & Singer report that most of these axons degenerate. They also made observations on resected segments of the crayfish opener bundle. The large (presumably motor) axons did not show morphological signs of degeneration for the term of their experimental observations – 35 days. Thus it is not likely that enucleated crayfish axons are maintained by nucleated partners running in the same nerve trunk. Furthermore this clearly cannot be the case in sectioned third roots of the crayfish abdomen that innervate the flexor musculature, because they are purely motor and contain no sensory (hence nucleated) fibres.

Maintenance of enucleated axons

This brings us back to how soma-less crayfish axons can persist for long periods of time. The overriding question in crayfish is the source of metabolic input for enucleated axons. These axons are not in a 'resting state'; the metabolic load on them is considerable. Consider the case of the claw opener axon. In the first place it innervates two different sets of muscles, the stretcher of the propodite and the opener of the dactyl segment. Bittner (1968) has made an estimate of the innervation load of this axon: there are about 300 muscle fibres in the opener muscle and about 1000 in the stretcher. Because crustacean motor neurons form terminals about 60 μm apart on each muscle fibre (multiterminal innervation) each muscle fibre carries about 50 terminals; this amounts to about

65000 total synapses for the opener motor axon. As was pointed out above, miniature JPs could be recorded from either opener or stretcher fibres, even involving axons that had been enucleated over 100 days. Miniature JPs represent a loss of transmitter from the presynaptic terminals which must presumably be stored or constantly resynthesised (or it could be recycled by uptake back into the terminals, which itself might be a costly metabolic process if uptake were active).

It is clear that the semi-functional state of sectioned crayfish axons places heavy demands upon the synthetic capacities of these stumps. There are three possibilities: the axons are able to maintain themselves, the glia associated with them provide metabolic 'inputs', and finally, substances pass in a retrograde direction, from muscle fibre, through the synaptic cleft, and into the terminal regions of the axon. Subsistence through materials supplied from the muscle is not likely because it was shown that a resected nerve, a segment that is isolated from *both* soma and peripheral connections (opener muscle fibres), showed no degeneration of (presumably) motor axons (Nordlander & Singer, 1972). It is possible that the axon is carrying on with axonal constituents that were provided by the soma before the nerve was sectioned. If the cell body does contribute substances critical for cellular function of axons, they have a very long half-life indeed. It seems more likely that although contributions from the soma are ultimately critical (crayfish axons do eventually degenerate) they do not need to be continually replenished, or can be provided from other sources. Thus while the soma might be sufficient for maintaining the health of its axon, its *continual* presence is not necessary. One of the first reactions to axon section consistently noticed in the early experiments was that of glial hypertrophy; glia were thus suspected to play a key role in nourishing soma-less axons (Hoy, 1968, 1969). Nordlander & Singer (1972) noticed that within 48 h after nerve section, glial layers around injured axons began to swell, being most noticeable in the layer adjacent to the axon, the adaxonal layer. This swelling and accumulation of organelles such as mitochondria, ribosomes, rough endoplasmic reticulum, and vesicles, a condition which persists for months, seems to favour the hypothesis that glia are playing a supportive role in maintaining the distal segment of sectioned axons.

Regeneration in crayfish

In the crayfish neuromuscular systems so far studied, regeneration occurred rapidly, sometimes within a few weeks, and resulted in restoration of normal function as interpreted at the level of behaviour and synaptic electrophysiology. From physiological data the most likely mechanism of regeneration, consistent with parsimony, is that of fusion of the outgrowing proximal axon stump with its still-persistent homologue in the distal stump. In the only published electron microscopic study of crayfish regeneration to date, Nordlander & Singer (1972) failed to find evidence of fusion between outgrowing sprouts from the proximal axon and axons in the distal stump. They found numerous sprouts growing into the distal trunk and interpret their findings in terms of the conventional outgrowth mechanism of regeneration. It is true that their results are consistent with the outgrowth and re-innervation mechanism, but they do not exclude regeneration by fusion. Fusion also requires proximo-distal outgrowth and our notion of fusion, based on physiological evidence, does not limit the extent of outgrowth into the distal segment. Specifically, fusion need not be interpreted as an abuttment of severed central and peripheral stump ends. It is more likely, in my opinion, that fusion occurs over a substantial length of overlapping stumps (sprouts) of the outgrowing proximal axon and the persistent distal homologue. The main difference between regeneration by fusion and outgrowth is that fusion does not require the establishment of new neuromuscular junctions with the muscle, whereas outgrowth does. Nordlander & Singer followed regenerating axon sprouts as far as 1 to 1.5 cm distal to the lesion site. To determine whether or not fusion has occurred it might have been necessary to follow central sprouts quite far peripherally, possibly even as far as the junctions.

Although it is almost certain that the seven 'giant' axons studied by Nordlander & Singer were motor axons, and that one of them was indeed the opener excitor, positive identification was not made. Further studies, in which electrophysiology, electron microscopy, and behaviour are combined, might be necessary in order to confirm, or deny, the occurrence of axonal fusion.

It is important to recall that the innervation load for the claw opener excitor axon is about 65 000 synapses. In the claw, regeneration occurred in some cases as early as two weeks. In such instances the muscle did not seem to be innervated sparsely, in a few fibres, as might be expected from regeneration by outgrowth, but in many. This too, is consistent with regeneration by fusion.

Neural regeneration in insects

This topic is considered by Young and by Edwards (this volume) so only a few remarks will be made for comparison with crayfish. The studies most comparable to those in crayfish were done in the cockroach and locust. If we restrict ourselves to motor systems and central connectives, differences are quite apparent between these closely related insects. There seems to be no question that regeneration in sectioned motor nerves to the walking legs in cockroaches occurs by outgrowth and re-innervation of the muscle (Bodenstein, 1957; Guthrie, 1962, 1967; Jacklet & Cohen, 1967. Regeneration of axons in the central connectives of the locust, *Schistocerca*, was studied by Boulton (1969) using electron microscopy. Regenerating nerve sprouts began to grow out from proximal stumps within a week of axon section, and this observation coupled with the absence of degeneration in distal stumps led to the conclusion that regeneration was occurring by axonal fusion, as proposed in crayfish. Hence again, locusts seem to more resemble crayfish than cockroaches, which in turn are more like vertebrates, in the behaviour of axons to transection. Evidence for axonal fusion has thus been presented for two arthropod species, crayfish and locust.

Conclusions and Summary

There can be no question that enucleated axons of crayfish motor neurons can survive without degeneration for many months in the absence of metabolic contributions from their nucleated cell bodies. Ultimately, crayfish axons do degenerate; the time required to do so, however, raises fundamental questions about the source of their metabolic maintenance. Whatever the nature of the day-to-day contribution of cell bodies, it is compensated for by constituents either stored or synthesised within isolated axons themselves and/or associated neuroglial cells, which are nucleated. The ability of axons to persist in the absence of nuclear connections is also known in parts of the locust nervous system, although the time scale of survival is less striking than in crayfish. The problem of maintenance of enucleated axons in locusts seems more a matter of inhibiting phagocytosis than of a possible 'trophic' relationship between axons and glia. While the mechanism(s) that inhibit phagocytosis in crayfish axons are not clear, the ability to persist in a soma-less state for over half a year forces one to focus attention upon the source of metabolic maintenance for sectioned axons, and neuroglia seem to be the leading candidate.

The rapidity of regeneration in crayfish motor nerves, months before degeneration of the distal stump would be expected, makes questionable the applicability of the proximo-distal outgrowth type of regeneration that is typical of vertebrates and some invertebrates. Instead, physiological evidence supports regeneration by reconnection of distal and proximal axons by fusion. While one ultrastructural study (Nordlander & Singer, 1972) did not find evidence for fusion, their observations do not exclude it. Further studies combining electron microscopy and electrophysiology are warranted, especially in view of the discovery of hyperinnervated muscles (cited in this paper, above), to determine by what mechanism(s) regeneration can occur.

By whatever mechanism regeneration in crayfish motor axons occurs, the result not only restores behaviour but also the specificity of the synaptic properties of regenerated neurons. When these studies were initially conducted, it was difficult to know whether regeneration resulted in a precise reconnection of central or peripheral portions of a given neuron. With the development of new neuron mapping techniques, it should be possible to investigate the question of the specificity of regeneration, interpreted at the level of uniquely identified, single neurons (see Young, this volume).

It is to be hoped that the well known model systems of arthropod neural networks will prove as illuminating in such an important area of developmental neurobiology as neural regeneration as they have been in understanding the neuronal basis of behaviour.

Much of the experimental work presented in this paper was done in collaboration (at Stanford University) with Drs George D. Bittner and Jan Bruner and has been, or will be, published elsewhere. Dr Jeffrey Wine generously shared his unpublished results, and along with Ms Carol Mason, kindly criticised the present manuscript. During the preparation of this paper, discussions with Drs D. Kennedy, C. H. F. Rowell, and H. Bern were very helpful and are here acknowledged.

References

Birks, R., Katz, B. & Miledi, R. (1960). Physiological and structural changes at the amphibian myoneural junction in the course of nerve degeneration. *Journal of Physiology, London*, **150**, 145–68.

Bittner, G. D. (1968). Differentiation of nerve terminals in the crayfish opener muscle and its functional significance. *Journal of General Physiology*, **51**, 731–58.

Bodenstein, D. (1957). Studies on nerve regeneration in *Periplaneta americana*. *Journal of Experimental Zoology*, **136**, 89–115.

Boulton, R. S. (1969). Degeneration and regeneration in the insect central nervous system. I. *Zeitschrift für Zellforschung und Mikroskopische Anatomie*, **101**, 98–118.

Boulton, R. S. & Rowell, C. H. F. (1969). Degeneration and regeneration in the insect central nervous system. II. *Zeitschrift für Zellforschung und Mikroskopische Anatomie*, **101**, 119–34.

Bruner, J. & Kennedy, D. (1970). Habituation: occurrence at a neuromuscular junction. *Science*, **169**, 92–4.

Bruner, J., LeDouarin, G. & Hoy, R. R. (1971). La transmission neuromusculaire après gangliectomie chez l'Écrevisse. *Journal de Physiologie*, **63**, 178–9.

Farley, R. D. & Milburn, N. S. (1969). Structure and function of the giant fibre system in the cockroach, *Periplaneta americana*. *Journal of Insect Physiology*, **15**, 457–76.

Girardier, L., Reuben, J. P. & Grundfest, H. (1962). Changes in membrane properties of crayfish muscle fibers caused by denervation. *Federation Proceedings*, **21**, 357.

Grafstein, B. (1969). Axonal transport: communication between soma and synapse. In *Advances in Biochemical Psychopharmacology*, ed. E. Costa & P. Greengard, vol. 1, pp. 11–25. New York: Raven Press.

Guthrie, D. M. (1962). Regenerative growth in insect nerve axons. *Journal of Insect Physiology*, **8**, 79–92.

Guthrie, D. M. (1967). The regeneration of motor axons in an insect. *Journal of Insect Physiology*, **13**, 1593–611.

Hess, A. (1960). The fine structure of degenerating nerve fibers, their sheaths, and their terminations in the central nerve cord of the cockroach (*Periplaneta americana*). *Journal of Biophysical and Biochemical Cytology*, **7**, 339–44.

Hoy, R. R. (1968). Regeneration and degeneration in the nervous system of the crayfish, *Procambarus clarkii*. Doctoral Dissertation, Stanford University.

(1969). Degeneration and regeneration in abdominal flexor motor neurons in the crayfish. *Journal of Experimental Zoology*, **172**, 219–32.

Hoy, R. R., Bittner, G. D. & Kennedy, D. (1967). Regeneration in crustacean motoneurons: evidence for axonal fusion. *Science*, **156**, 251–2.

Jacklet, J. W. & Cohen, M. J. (1967). Nerve regeneration: correlation of electrical, histological, and behavioral events. *Science*, **156**, 1640–3.

Johnson, G. E. (1926). Studies on the functions of the giant nerve fibers of crustaceans, with special reference to *Cambarus* and *Palaemonetes*. *Journal of Comparative Neurology*, **42**, 19–33.

Kennedy, D. & Takeda, K. (1965*a*). Reflex control of abdominal flexor muscles in the crayfish. I. The twitch system. *Journal of Experimental Biology*, **43**, 211–27.

(1965*b*). Reflex control of abdominal flexor muscles in the crayfish. II. The tonic system. *Journal of Experimental Biology*, **43**, 229–46.

Milburn, N. S. (1971). Neurosecretion in endings of motoneurons during limb regeneration in cockroaches. *American Zoologist*, **11**, 673.

Nordlander, R. H. & Singer, M. (1972). Electron microscopy of severed motor fibers in the crayfish. *Zeitschrift für Zellforschung und Mikroskopische Anatomie*, **126**, 157–81.

Roeder, K. D. & Weiant, E. A. (1950). The electrical and mechanical events of neuromuscular transmission in the cockroach, *Periplaneta americana*. *Journal of Experimental Biology*, **27**, 1–13.

Rowell, C. H. F. & Dorey, A. E. (1967). The number and size of axons in the thoracic connectives of the desert locust *Schistocerca gregaria* Forsk. *Zeitschrift für Zellforschung und Mikroskopische Anatomie*, **83**, 288–94.

Titeca, J. (1935). Étude des modifications fonctionnelles du nerf au cours de sa dégénérescence wallérienne. *Archives Internationales de Physiologie*, **41**, 1–56.

Tung, A. S. C. & Pipa, R. L. (1971). Fine structure of transected interganglionic connectives and degenerating axons of wax moth larvae. *Journal of Ultrastructure Research*, **36**, 694–707.

Usherwood, P. N. R. (1963*a*). Response of insect muscle to denervation. I. Resting potential changes. *Journal of Insect Physiology*, **9**, 247–55.

(1963*b*). Response of insect muscle to denervation. II. Changes in neuromuscular transmission. *Journal of Insect Physiology*, **9**, 811–25.

Usherwood, P. N. R., Cochrane, D. G. & Rees, D. (1968). Changes in structural, physiological, and pharmacological properties of insect excitatory nerve–muscle synapses after motor nerve section. *Nature, London*, **218**, 589–91.

Weiss, P. & Hiscoe, H. B. (1948). Experiments on the mechanism of nerve growth. *Journal of Experimental Zoology*, **107**, 315–95.

Wiersma, C. A. G. (1960). The neuromuscular system. In *The Physiology of Crustacea*, ed. T. H. Waterman, vol. 2, pp. 191–240. New York: Academic Press.

Neuron constancy and connection patterns in functional and growth studies

G. A. Horridge

Introduction

Anatomical identity in functional studies of neurons

Before I was born it was known from the work of Cajal and others that nervous systems of all animals are fantastically complex assemblies of neurons, and that their obvious function is to connect with other distant neurons and interact in very precise but unknown ways. The basic idea that each axon is a line along which impulses speed was established by the end of the nineteenth century, and the main attributes of reflex action had then been described by Sherrington and his associates. Understanding how nervous systems control behaviour depended upon the 'black box' picture of a *synapse*. The physiological properties of a synapse were not directly known, and the fine structural detail to which we are now accustomed was replaced by a convenient 'gap' between the neurons. The explanatory model synapse and axon pathway combined whatever attributes were wanted for the explanation of function because the actual anatomy was incomplete and very complex. In brief, this was the climate of thought in the textbooks and reviews when I was having unrelated tutorials in neurophysiology and animal behaviour with other undergraduates. Our other reading and our midnight discussions, however, were often concerned with the limitations of knowledge set by available techniques, with the problem of designing experiments not based on the conclusions of our immediate predecessors, and by the inability of the classificatory system of our minds to analyse behaviour when the problem was to separate parts and processes that all the time were feeding back upon one another, not equally so and not linearly. One of the reasons that I persisted with coelenterates, although the new microelectrode techniques could not be applied to them, was that at the level of a nerve net it was then possible to include a larger fraction of the anatomical, physiological and behavioural findings in one explanatory diagram than could be considered for a higher animal with a chain of ganglia.

Of course, nowadays, even that illusion has dropped away, for it has become obvious that nerve cells and conducting epithelial layers in coelenterates are hardly understood at all, and their more complex behaviour cannot yet be approached by any explanation. My own efforts with simple nervous systems led me to advocate that physiology should be combined with elucidation of the actual anatomy of the neuronal interconnections at the cell level. The same realisation among others explains the revival of methylene blue, Golgi and silver staining techniques in the fifties and sixties. If we couldn't see the wiring diagram at least we could discover the distribution of the wires.

The new post-war physiological technique, the use of the capillary microelectrode which reveals the activity of some single nerve cells, has also been overriding in its effect. The euphoria, the promise of explanations of complex nervous activity, is there for all to read in the early papers of the fifties, perhaps best illustrated in books by Eccles. The graded summation of inhibitory and excitatory synaptic potentials could certainly explain many aspects of spinal reflexes and, it was hoped, could explain the whole of behaviour. But, more relevant to my present theme was the giant fibre and giant neuron bandwagon of the post-war period. Probed by membrane biophysicists for their own ends, a wide range of large neurons were explored. First, current was passed through their membranes and their bathing medium was changed with no respect for the normal physiological milieu. The interesting historical developments were not the tabulated data on membrane constants but the realisation that these overgrown nerve cells are individually identifiable and the growing importance of anatomy to physiologists. They began to talk about the small posterior cells, the medio-lateral giant fibre and the parabolic burster, and I shall return to this important change in outlook under the heading of identifiable cells.

Driven by the necessity to know the wiring diagram, physiologists showed renewed interest in neuroanatomy. Retzius, Zawarzin and Cajal came back into vogue; their works were searched for convenient large neurons. Perhaps it was this movement more than anything else which spurred on Ted Bullock and myself to attempt to bring all the old, invertebrate neurobiological literature back into circulation. Exactly at the same time Wiersma was showing by tedious extracellular recording that the central fibres of the crayfish can be uniquely identified, given functions in terms of input fields, and found in the same place in every individual of the species. However when I now look back at our book (Bullock & Horridge, 1965) I am struck by the casual way in which neuron constancy was presented as constancy of cell bodies, constancy of axon destination, constancy of neuron function where identified, and unknown variability of anatomical synaptic wiring diagrams. The classical vertebrate neuron anatomists had always accepted constant neuron *types*; others working on annelids, molluscs and arthropods had described the constancy of large neurons with little comment, evidently taking for granted a corresponding functional constancy.

At the time I was searching for alternative unorthodox systems that were outlined in a paper that drew critical but uninformed comments. In the early sixties we were all rather disturbed at the fantastic numbers of fine fibres which had been revealed by electron microscopy of neuropile and the difficulty of identifying synapses with their neurons of origin. It was possible to say truthfully that 'From anatomical studies of neuropiles no-one has ever obtained anything that looks like a wiring diagram' (Horridge, 1961). Few, however, seemed interested in the possibility that a complex pattern of synaptic pathways could be formed by the specificity of transmitters in a tangle of naked neuron processes. But this possibility has still not been ruled out, especially as a mechanism of interaction during minutes or hours rather than milliseconds, and as a normal mechanism of hormonal effects on central neurons. Significantly, in the example of neuropile known in most anatomical detail, the lamina of the fly optic lobe, it is possible that each cartridge of about a dozen tightly packed neurons operates by a mixture of constant-location anatomical synapses, anatomically indefinable sensitivity to particular transmitters released outside localised sites, and a flow of current dependent on the geometry of the system. In fact the best known cases prove that an anatomical analysis in isolation must wait upon the physiological analysis for even the most superficial guess as to function.

The physiological analysis of single neurons, therefore, emphasised the importance of identifiable cells, and the last few years have seen progressively greater use of anatomical techniques by electrophysiologists to this end; until now it is hardly respectable to record from unknown 'units' without relating them to anatomical neurons as far as possible. Injection of neurons with dyes through the electrode, and subsequently with cobalt ions, means that neurons from which recordings are taken can then be visualised, and their part in the pattern of connections can be inferred in more detail. We can then argue from the pattern of possible anatomical contacts and the known physiological interactions towards something that may approach the real pattern of functional connections between the neurons. Even so, every bit of the map is tentative until repeatedly triangulated from several different points of view.

For the physiologist, the consequence of working with identified neurons is that he can describe the properties of *this* motor neuron as distinct from *a* motor neuron. At the lower limit, he may know the pattern of anatomical synapses upon the cell but he has difficulty in distinguishing *this* synapse from *that* upon a single cell. However, because he knows what neuron he is talking about he can examine physiological connections between known cells and can describe their idiosyncrasies. The point is that analysis of this kind can proceed as far as the ability of the experimenter to distinguish the neurons by criteria other than those he uses for his own technique. He is limited to discussion of physiological classes of units if that is what his techniques, and the material, reveal. Exactly in the same way, to draw a map of connections between neurons, knowing only classes of them, limits us to drawing diagrams about how typical members of a class may be linked with those of other classes. In most vertebrate systems the effort is still at this stage, so the need for identified neurons is not felt. However, labelling individual neurons is the first step towards finding the functionally effective anatomical connections between them *as individuals*, and, since the neuron is the unit of the nervous system, it is commendable to try to find out if the basis of behaviour really lies in the pattern of connections between neurons.

Determination and effects of neuron size

In many nerves and axon bundles it has long been known that the larger axons have the lowest threshold in *electrical* stimulation, the fastest conduction velocity and usually the most phasic properties (Bullock & Horridge, 1965 review, Chapt. 5). In several studies of vertebrate motor neurons it has more recently been shown that the larger ones innervate greater numbers of muscle fibres. The number of synaptic potentials required to discharge a motor neuron, the amount of transmitter it releases, its mean rate of firing and its rate of protein synthesis, are all directly related to its *size* (Henneman, Somjen & Carpenter, 1965*a*, *b*; Peterson, 1966). On the sensory side, it has been found that the larger the sensory fibre, the greater is the excitatory effect of its impulses on post-synaptic cells (Mendell & Henneman, 1971). The order in which motor neurons are brought into play by increasing frequency and number of descending command signals is similar to the order in which they are silenced by proprioceptive feedback, and is also determined mainly by their size. Neuron size, therefore, has functional significance in vertebrates.

Among invertebrates, giant neurons and identified large cells have long been recognised, but in fact we know little about the significance of their size differences and nothing about the control of size. Giant fibres are characteristic of startle systems. Giant interneurons activate large parts of the body via numerous other neurons; giant motor neurons excite many muscle fibres. Large size is probably an adaptation to the quantity of transmitter they must secrete, and carries the added advantage of speed of conduction. Even so, no reasonable explanation is forthcoming for the existence of a few huge neurons (over 500 μm) and several large neurons (over 200 μm) in ganglia of many slow-moving molluscs.

For lobster motor neurons in the range 25–125 μm diameter, the small motor neurons cause small facilitating EJPs in the muscle fibres, whereas the large ones cause antifacilitating large EJPs. The small motor neurons are excited more easily by central excitation and are recruited more easily by continued interneuron firing, and small motor neuron firing is less easily adapted to a constant presynaptic interneuron stimulus, whereas large motor neurons adapt quickly (Davis, 1971). Indications are that similar correlations of functional properties with neuron size will be found throughout the nervous system in many groups of animals. Thus, in many ways, properties correlated with neuron size participate in the orderly control of a reflex or a centrally controlled movement.

The only recent discussion of factors controlling neuron size illustrates how little is known (Jacobson, 1970). Neurons which do not divide in the life of the animal grow as the animal grows, but not in proportion. The largest neurons are those formed earliest in development; which may be significant, or only a consequence of the ability of the observer to recognise them earlier than their smaller neighbours.

Establishment of connections: vertebrate systems

Almost nothing is known of the mechanisms of establishment of neuron connections in vertebrate embryos. By generalisation from observation of living neurons in tissue cultures and transparent embryos, we believe that each fibre tip advances by putting out numerous fine pseudopodia that extend and retract in the tissue ahead. Observation of the final pattern is a poor indication of the mechanisms by which particular pseudopods survive while others do not. It is supposed that the large numbers of embryonic neurons which die do so because they fail to make adequate connections. From recent reviews (Horridge, 1968*a*; Edds, Barkley & Farnbrough, 1972) a few generalisations emerge. The major classes of neurons are genetically differentiated and clearly make connections with the appropriate class of target cell. There are several instances of a powerful mechanism whereby a cell, once innervated, will not accept another terminal upon it, e.g. some mammalian muscle fibres. There are several authentic cases of continual death of terminals and growth of replacements, e.g. in taste buds. Any disturbance can cause adjacent neurons to sprout and branch, such changes are typical of metamorphosis and presumably of development in general. A shift of terminals during development is known in the optic tract contralateral projection of *Xenopus* and is inferred for the ipsilateral projection, so that the fields of the two eyes remain in register on each half of the optic tectum. The major framework of connections is thought to be laid down independently of whether it functions, but there is a general expectation that details are significantly influenced by function at critical stages in development. Several lines of evidence from optic projections, sympathetic ganglion regeneration and motor axon regeneration suggest 'that the invading fibres compete upon a background of preferences of the cells that they innervate' (Horridge, 1968*a*, p. 306). Putting this to its widest level, it would be interesting to know whether a balanced input of natural stimuli during maturation is essential for the ability of an animal to appreciate its normal share of natural variety.

Regeneration of motor axons in lower vertebrates

In higher vertebrates, including man, muscles that are re-innervated by a cut nerve can no longer be co-ordinated properly. The inference from many studies is that whatever specificities may have been laid down in development they are swamped by loss of specificity in either the axons or muscles or both, because the motor axons grow back indiscriminately to the muscles. In newts, salamanders and teleost fish, however, function is well restored in a few weeks. Recent experiments suggest that the explanation does not lie in the changed function of the motor axons but in the ability of the different motor axons to branch and spread until they meet their correct muscle partners.

In one of these lower vertebrates, when a muscle is deprived of a nerve it will accept innervation by incorrect axons. An example is in the innervation of the goldfish superior oblique muscle by motor axons normally running to the inferior oblique muscle. When the original motor axons of the superior muscle are now allowed to return, a very interesting thing happens. The synapses of the fibres normally running to the inferior oblique muscle persist structurally on the superior oblique but become physiologically ineffective (Mark, Marotte & Johnstone, 1970).

Recently Mark (personal communication) has found that in the axolotl the regional innervations of leg motor axons overlap. There are structurally perfect, but non-functional motor synapses where a muscle fibre is innervated in the normal course of events by the wrong axon, as well as by the correct one. One can now see that the multi-terminal polyneuronal innervation of the muscles in lower vertebrates by widely ramifying branches of motor axons requires a rigid mechanism whereby the axons connect with their appropriate muscles, and if mistakes are made the unwanted synapses can be switched off. Of course, it is of great interest whether this mechanism persists in the central nervous system, and whether it has been taken over as a mechanism of memory.

Xenopus visual system

The original observation by Sperry was that the axons of a cut Salamander optic nerve regenerate their former projection to the optic tectum (and perhaps elsewhere) in the brain. The new topotopical projection was subsequently demonstrated by stimulating the retina point by point and recording evoked potentials. Sperry suggested that the regenerated terminals found their proper target cells by a uniquely differentiated chemical affinity. Sperry's observations did not, in fact, justify a theory in terms of differences between individual cells, because he was not working at this fine level. The subsequent work with evoked potentials showed only polarity, and a region by region accuracy in the distribution of the terminals.

Then it was discovered in *Xenopus* that axons growing (for the first time) from *half* a retina come to occupy positions in proper sequence across the *whole* tectum. Subsequently it has been found that normal growth of the tectum is from front to back and lateral to medial (i.e. from one corner) up to about stage 66 (metamorphosis). Therefore the projection shifts its position as development proceeds. Also, as growth occurs, the projection improves in detail by reduction in the sizes of the fields of units (Gaze, Chung & Keating, 1972).

Discussing the findings, the authors state 'The thesis that specific chemical affinities govern the formation of neuronal connexions *implies a fixed synaptic relationship* between a given retinal ganglion cell and its tectal counterpart(s).' I would argue that a changing pattern of specific regional affinities could also determine a changing distribution of terminals. However, they point out that the observed shift in the projection during development and its unusual location in this special case of regeneration cannot be explained by any rigid system of uniquely matching retinal and tectal cells, as proposed in the original formulation of the hypothesis of neuronal specificity. Ambiguously they say that the retinal ganglion cells possess innately acquired differences from one another, and that the differences manifest themselves by the different sites at which they settle. Then they conclude that the retinal terminals link up with tectal cells under the control of factors which appear to be operating on a system-to-system rather than a cell-to-cell basis.

Considering this from an entirely different background leads me to the following comments,

(*a*) A wide range of data suggests that *neuron classes* of vertebrates, and many individual neurons of invertebrates, do differentiate from each other, and that their chemical differences are the only imaginable basis for the determination of the broader outlines of neuron size, distribution and synapse formation.

(*b*) At the same time, regional differences caused in unknown ways by temporal differences and spatial separation are common in studies of development of other cells.

(*c*) Results so far on the first projection are not relevant to the formation of connections because target cells are not distinguished by an independent criterion. The only information on choice of target cells comes from the ipsilateral projection.

(*d*) The constantly changing pattern in development must be the result of parallel changes in tectal and retinal cells, and this could be based on chemical differences between cells. One gradient in time or space is an inadequate explanation.

(*e*) Where there is growth of half a retina into a whole tectum, the vacuum created by the experiment may be causing de-differentiation and restructuring in the tectum; it is not necessarily lateral interaction between the incoming fibres. One gradient of lateral interaction is again an inadequate explanation.

(*f*) When a two-dimensional array projects on another two-dimensional array, any observation is on the *relationship* between the two, and it is not easy to decide whether one or both have been changed by the experiment or by passage of time.

(*g*) Perhaps the most telling point is that although a variety of transplantations and observations have been made over years of careful work by a dedicated and talented team, the major result is to reject fixed cell-by-cell chemical differences and replace them by a vague appreciation that retinal axons are distributed as a system on the tectum as a system.

In chickens and in some fish, when most of the retina is destroyed and the optic fibres of the remainder are allowed to regenerate, axons grow to the correct region of the tectum, and do not spread into vacant areas. So, one has to decide whether to keep separate the observations on different species or to be more general by considering data from several. At present, results from one species contradict theories derived from another.

Other relevant findings in vertebrate CNS

Projections of numerous fibres in parallel from one centre to another are a conspicuous feature of the vertebrate central nervous system. Ingredients in the vertebrate success story are, first, axons in parallel paths from one class of neurons can terminate at locations that maintain the spatial relations of neurons upstream on the excitation pathway, and secondly, they have been forced to evolve mechanisms whereby this process can depend on function because there are too many neurons for each to be specified in every detail.

Ascending fibres of the rat spinal cord project the body surface upon the medulla nucleus. When axons from hindleg areas are removed, the projection of the forelegs spreads, presumably by sprouting of additional terminals but there is little evidence of an actual migration from former positions (Wall & Egger, 1971).

In an elegant series of experiments, Gaze, Keating, Székely & Beazley (1970) have found that, in the projection of the two eyes to each side of the optic tectum in *Xenopus*, the (indirect) ipsilateral projection lies superimposed upon the (direct) contralateral projection, so that visual fields of the two eyes are in register on each side of the tectum. Efforts to disturb this alignment by eye rotations and similar operations in early life are followed by a recovery that can only be interpreted as dependent on function. Some critical factor is essential, for an independent attempt to confirm the findings was unsuccessful (Jacobson, 1971).

Work on vertebrate neural projections is concerned with the gross distribution of terminals, and the regional differences between neurons within a class. There is, as yet, no sign of analyses of an identified circuit, of synapse formation, or of particular target cells.

Two other areas of investigation of formation of circuits have recently emerged. First, techniques are now available to modify the abundance of units with known properties in the cat visual system by allowing a kitten to see only one pattern during a restricted visual experience at a critical period in its early life. Kittens seeing only vertical stripes develop a preponderance of cortical units sensitive to movement of a vertical edge (Hirsch & Spinelli, 1970; Blakemore & Cooper, 1970). It is not known whether the changed proportion is caused by disappearance of horizontally sensitive cells, or by their conversion, and clearly the difference is crucial.

Secondly, numerous examples are known of well-defined regional separation where several types of presynaptic fibre synapse upon one target neuron, e.g. the Mauthner fibre of fish, and upon the dendrites of hippocampal neurons in mammals. As in the projection work, there is a *relationship* between neurons of different classes, with spatial and temporal factors in addition, and none of the causal factors are yet guessed at.

Establishment of connections: invertebrate systems

Since the classical works of Cajal, Lenhossek, Retzius and others on the structure of the central nervous system of a variety of invertebrates, it has been recognised that the large neurons of typical invertebrate central ganglia are recognisable from animal to animal. This is particularly true of annelids, and arthropods. A given neuron always ramifies over the same region, with its cell body usually in a characteristic position. The pattern of the dendrites is recognisable for a given cell and different from cell to cell: the distination of the axon is a fixed feature of that neuron. This individuality of the large neurons has always been clear in the best anatomical studies of neuron types in these animals. Much of that work was done over seventy years ago but until recently physiological studies in the intervening period have not been deeply concerned with this fixed pattern of individually different neurons. Recently, however, there has been a renewal of interest, because, given an identifiable neuron of constant, complex form and with known relationships by functional synapses with other identifiable neurons, there is a hope that new principles and properties of the differentiation of unique cells will be discovered. The effects of operations, transplantations, stimulation, hormones, learning, pre- and postsynaptic cell loss, and factors controlling synaptic sites can all be studied on a neuron that can be found again and again in different animals.

The next two examples show that in some cases neurons can be substituted for other neurons of the same class which would normally connect with different target cells. They show that neurons which lie in different locations (but serve a similar function) can substitute for one another, but they illustrate a principle of greater significance, that tests relating to specificity depend on what can be done within the range of experimentally available situations.

Segmental repetition of neurons: the insect leg

Working with identified motor neurons, Young (1972) has examined the innervation of cockroach legs transplanted from one segment to another. The first step was to identify certain large motor neurons of the meso- and metathoracic ganglia which supply muscles of the leg and, by stimulating the cell bodies individually, to find their peripheral distribution. Cells marked by dye were constant in cell body position and peripheral distribution. The process was repeated in metathoracic legs transplanted to the mesothoracic segment. To reach their destinations motor axons of the insect leg have to grow past many other muscles. However, an identified central neuron which formerly sent its axon to a particular identified muscle in the second leg now sends its axon to the corresponding muscle in the third leg growing in the place of the second.

This observation shows that the mechanism by which an identified cell grows only to a given muscle is repeated (or approximated) in the next segment. Possibly there is a generalisation that the mechanism for the finding of all target cells is repeated segmentally; if so, the progressive diversification of this in cephalisation will be an interesting topic.

Repetition of neurons: the compound eye

In each ommatidium of the compound eye of the fly, six of the photoreceptor cells have axons which run to different synaptic regions of the lamina below. An eye with 2000 facets has 6000 receptor axons of this type, each running by its own distinct and particular pathway to the second-order cells. In one normal eye, every one of 600 axons which was examined ran to its correct destination (Horridge & Meinertzhagen, 1970).

The six axons under discussion from each ommatidium are clearly different, and they are repeated in each ommatidium across the eye. Although every axon takes a different path, it is unlikely, and unnecessary, to suppose that 6000 differently labelled neurons correspond to the 2000 synaptic sites of the lamina. However it is inadequate to suppose that only six different labels exist for the major types of axons to find their destinations. The first complication is that each synaptic target region of the lamina receives an axon from six different ommatidia. Therefore labels of second order cells cannot correspond directly with those of the first order cells. Secondly, the axons spread laterally in particular directions that are different for each of the six types.

A question of interest is whether there are regional differences across the layers, so that first order cells correspond with individual predetermined second order cells below them. Gradients could depend on spatial position or on time of development. Eyes that are transplanted to normal sites with 180° rotation in the locust will regenerate some vision (Horridge, 1968*b*). In a fly eye with abnormal arrangement of facets, the second-order cells can receive incorrect first order axons (Meinertzhagen, 1972). Therefore neurons are able to substitute for another within the same class, although they all emerge from normal growth with individually different connections. One of the very curious facts about the abnormally connected axons in the fly is that their directions of growth and length appear to be partially fixed by the nature of their cell of origin.

Deeper in the insect optic lobe are visual units with fields restricted to parts of the whole visual field. As in all parts of all nervous systems, some factor must have limited the arborisations of these units to form physiological pathways only within a given region. As candidate mechanisms we can propose time differences in growth, gradients of epidermal origin carried trophically down receptor axons, or predetermined intrinsic regional differences in the optic lobe. To go this far, even, is logically dangerous: mechanisms cannot be inferred from the final products, but only from the results of operative experiments to test candidate mechanisms directly. The formation of connections in the optic lobe is discussed at length by Meinertzhagen (this volume).

Growth of arthropod motor neurons into muscle

The whole question of how re-innervating motor fibres progressively invade muscle in arthropods, requires examination (see Young, this volume). In regeneration of walking legs of *Grapsus* about 20% of muscle fibres are innervated in the 2-week-old limb bud and they show only poorly facilitating responses. Three weeks later most muscle fibres are re-innervated with the usual variety of facilitating junctions (Govind, Atwood & Lang, 1972). It is not known whether synapses themselves mature as they age, or whether the more effective ones are those formed later. The regenerated synapses show that 'both the determination on a single postsynaptic cell, and sets of terminals as one goes from one postsynaptic cell to another, are very deterministic in nature' (Atwood & Bittner, 1971). This work suggests a whole range of problems concerning the determination of the properties of synapses by the same kinds of mechanisms which determine the choice of partners.

G. A. Horridge

Neural feedback loops

The problem

The careful study of feedback loops by anatomical or physiological methods is not only of functional interest, but provides information about the consequences of the rules of growth.

To establish a feedback loop, the neurons require the necessary information to close the loop. As in the establishment of any other connection, the orderly growth of a feedback loop necessarily gives the superficial appearance of classes of neurons making selections based upon chemically differentiated cellular differences. Before this can be inferred, however, other factors, such as mechanical following of fibres already *in situ*, differences in maturation time, and changes caused by the neuron's own progress in forming contacts, must be considered. A neuron, *A*, growing a process into a region of neurons of a different type could make contact at random with a neuron, *B*, and so modify it that *B* accepts no other partner. *B* now grows out an axon which could follow the *A* fibre by contact guidance, or alternatively find *A* at random and recognise it by a chemical difference passed on by the first contact. An entirely different explanation is that *A* and *B* have differentiated by some kind of mosaic development as predestined partners for each other, but to suppose this degree of complexity without an exhaustive search for simpler rules of growth is a council of despair.

The convergence of Ia muscle spindle fibres on homonymous motor neurons

One of the best known feedback loops of the vertebrate nervous system is the central projection of the sensory axons from muscle spindles upon the motor neurons which innervate the same muscle. The sensory cell bodies situated in the dorsal root ganglion send a central process along the dorsal column. Branches from this longitudinal fibre drop down at intervals to the region of motor neurons, but the ramifications of these collaterals are limited. Only recently the accuracy of the system has been demonstrated (Mendell & Henneman, 1971). The conclusion, demonstrated electrophysiologically, is that each *Ia* afferent fibre sends terminals to all or nearly all of the 300 motor neurons of the particular muscle employed (medial gastrocnemius of cat). Conversely each motor neuron receives terminals from all or nearly all of its approximately 60 *Ia* fibres. However the same *Ia* afferent fibres terminate also on motor neurons of an adjacent muscle in 37 out of 67 cases tested. Tests at neighbouring regions of the cord suggest that a *Ia* fibre has a greater probability of making a synaptic contact with heteronymous cells at the same level as the homonymous cells than with similar cells which are farther away.

The above results indicate rules of growth as follows. The ingrowing sensory fibres seek out motor cells and ensure complete coverage. They err by making, or allowing to persist, 'erroneous' additional connections with heteronymous motor neurons. The patterns of synaptic potentials suggest that a motor neuron is not excited at more than one location by a given *Ia* input, as if, like some muscle cells, motor neurons once connected become unavailable to a second terminal of a *Ia* fibre. It is impossible as yet to say whether the first clues for the individuality of the set of feedback loops derive from the motor or sensory neurons or from the muscle.

In mammals the general pattern of feedback loops, *as known at the present time*, is that a class of neurons project back to the same or a different class only in a broadly disseminated way. For example in the inhibitory feedback loop of Renshaw cells back upon motor neurons in the ventral horn of the spinal cord, there is mutual inhibition of Renshaw cells, not only between those excited by motor neurons innervating different muscle groups, but also within the pool of Renshaw cells excited by motor neurons innervating single functional muscle groups (Ryall, Piercey & Polosa, 1971). The Renshaw cells themselves appear to project upon those motor neurons in their region of the cord that are related only in a general way to the motor neurons by which they are excited.

A feedback loop exact at single neuron level

The projection of the receptors upon the second-order lamina monopolar neurons of the fly eye is the best example of a connectivity pattern that has a functional significance and where every neuron meets exactly its own distinct target cell (Horridge & Meinertzhagen, 1970). The continuation into the optic lobe provides the only known case of a feedback loop where the feedback is upon a particular single cell and not diffusely upon a class of cells.

Between the lamina and the next neuropile layer, the medulla,

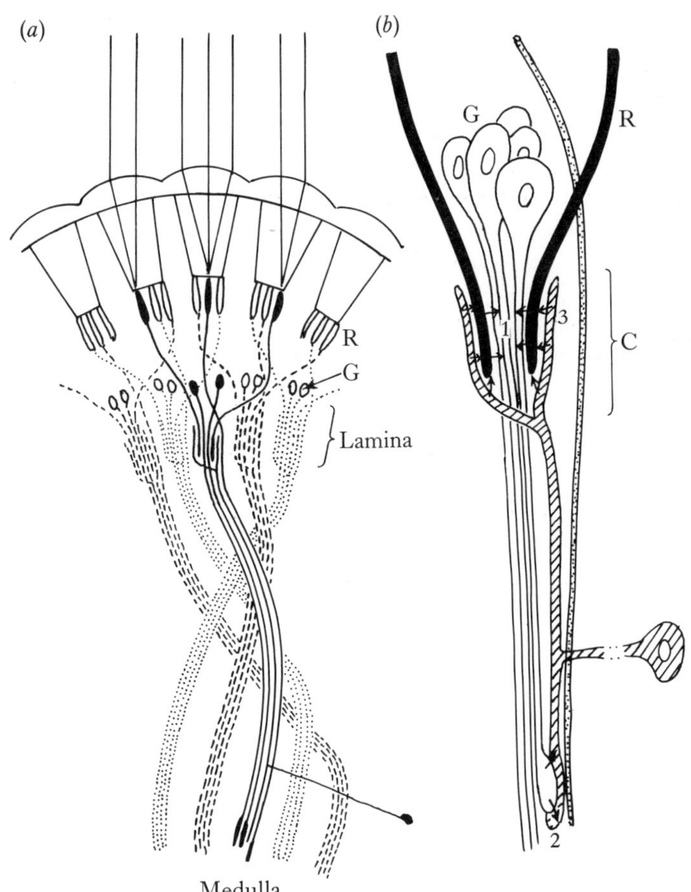

Fig. 1. The paths of individual axons in the 1st and 2nd projection of the fly eye.

(*a*) Parallel rays falling on a pattern of 7 facets (3 shown here) excite different receptors, as shown in solid black line. Receptors (R) converge upon ganglion cells (G) of the lamina in such a way that those excited by parallel rays come together in a single cartridge (C) of the lamina. Axons in the chiasma between the lamina cartridge and the corresponding medulla cartridges do not cross between bundles. Two of the receptor axons (stippled in *b*) bypass the lamina cartridge.

(*b*) The following synapses are known morphologically. 1, From receptor terminals (solid black) to lamina ganglion cells (white). 2, From lamina ganglion cell terminals in the medulla to fibres which also have terminals in the lamina (cross-hatched). These are the T or C cells of Strausfeld. 3, From the cross-hatched neuron back to receptor terminals of the same cartridge.

Although the cross-hatched fibre appears to form an angle-conserving feed-back loop, the arrangement is inferred from anatomical studies only. All fibres shown could possibly turn out to be afferent (inferred from Trujillo-Cenóz & Melamed, 1970; Horridge & Meinertzhagen, 1970; Strausfeld, 1971).

is a chiasma in which axons from the anterior of the lamina are transposed in a perfectly reversed array, bundle by bundle, upon the medulla (Fig. 1). Each bundle of axons running between a lamina cartridge and a medulla cartridge contains three main types of axon: (*a*) long retinula cells that run direct from retina to medulla, two from one ommatidium in each bundle; (*b*) lamina ganglion cells, five individually distinct ones in each lamina cartridge; and (*c*) several so-called 'efferent' fibres, which have terminals in lamina and medulla (Strausfeld, 1971).

The evidence that the feedback loop returns to the same neuron is as follows. In tracing the six largest axons of each bundle Meinertzhagen found no fibre crossed between bundles. Similarly, in Golgi preparations with two or three fibres staining by chance in a bundle, Strausfeld found that no fibre strayed from its partners. The terminals of a centrifugal fibre are presynaptic to the terminals of short retinula axons, and the terminal of a lamina monopolar cell makes contact with, and could be presynaptic to, the central ending of the centrifugal fibre (Trujillo-Cenóz & Melamed, 1970). Therefore there is considerable but not completely watertight evidence for the circuit illustrated in Fig. 1. Dangers in interpretation are, as usual, many. This feedback circuit may have a negligible effect compared with other neuron interactions; it is not known to be positive or negative feedback. Whatever its purpose the circuit evidently conserves the angular projection of the excitation, but its interest here lies in how it can be formed.

Here is a case where one might immediately infer a transfer of chemical specificity, but probably there has been a temporal sequence of growth of cartridges in succession, and only a brief period available for effective contact, during which the path of growth was very short and only one option was open. Transplant or regeneration experiments could therefore fail to reveal inter-loop differences (see also Meinertzhagen, this volume).

Identified cells

An individually identified neuron is preferable to one known only as a member of a class, because the neuron is the natural unit of the nervous system. To know which of them one is talking about is elegant and accurate. To take averages which smear underlying differences between individual cells is an unfortunate consequence of dealing with classes.

The ability to distinguish individuals or subclasses among neurons and their target cells is also most relevant to the investigation of patterns of connections and of how connections are formed. To say that a particular neuron forms a connection with its appropriate target cell implies that we have an independent means of recognising which is the appropriate target cell. To test the hypothesis that a particular neuron and its target cell are uniquely differentiated so that they connect with no others, we have to shift the target cell relative to the growing neuron to see whether this unique union is independent of place and time, and when other choices of partner are offered. Therefore we have to be able to identify both cells after the operation by quite other criteria. If only classes of cells can be identified, the results refer to classes, not individuals.

Limits of recognition of identified neurons

The division of a population into classes and subclasses is a familiar process that makes use of whatever distinguishing features are available. We can compare the neuron types of a vertebrate to species of trees, where we expect differences in branching pattern within a given type, but the larger neurons of an annelid, arthropod or mollusc are more like persons, where every one is recognisably different and twins come as a surprise. Axon destination, position of cell body, form of dendritic tree, transmitter content, appearance of granules within, membrane properties and correlation with behaviour are all relevant characters.

The whole concept of constancy of identified neurons depends on techniques, however, because those not yet shown to be identifiable apparently only await a technique to reveal it. In the above invertebrate phyla, where a technique is available to pick out a neuron in one individual, the distinct feature has always been found to be constant in the species (apart from obvious cases such as changes during a natural rhythm). For most groups of invertebrates, however, notably the Coelenterata, Ctenophora, Platyhelminthes, three lophophorate phyla, Nemertinea, three pseudocoelomate phyla, sipunculoids and related phyla, Echinodermata, Chaetognatha, Pogonophora, Hemichordata and Tunicata, the neurons are hardly yet known as major classes. In these groups neurons are small, very numerous and not likely to be distinguishable by the above characters. Moreover, giant fibres, which are a clue that at least some neurons are identifiable, are uncommon in these groups.

In the other groups, notably Annelida, Arthropoda and Mollusca, there are some neurons which are also numerous, small and ill-defined as classes. These phyla have concentrations of so-called globuli cells for which it is unlikely that any technique can demonstrate individual uniqueness in one animal, and much less a constancy between animals of the same species. All of the present concern is about a small proportion of large neurons which can be recognised and which are constant in different individuals or even in related species.

Already we have signs that the insect nervous system is not entirely defined down to the last dendritic twig and synapse location. In the optic lobe the primary receptors are fantastically regular in fly, cockroach, bugs and many others. Lamina structure shows that second-order and some of the third order neurons can be correctly categorised as individuals. In the second optic neuropile, the medulla, the neurons, which conserve the angle of the visual excitation can still be categorised consistently with the idea of fixity at cell level. The small intrinsic neurons of the medulla, however, and even more of the corpora pedunculata of the brain, cannot be given unique status, nor are they identical members of a class.

Even in insects we are obliged to accept the following classification on technical grounds:

(*a*) *Individually unique neurons*
- Motor neurons
- Isolated sensory cells
- Large interneurons

(*b*) *Repeated neurons with little variation*
- Compound eye retinula cells
- Optic lobe classes of larger neurons
- Grouped mechanoreceptor cells

(*c*) *Repeated neurons with relatively large variation*
- Small interneurons
- Globuli cells of corpora pedunculata
- Olfactory circuits

Fundamental difficulty of identity

A fundamental philosophical problem is the absence of any test for identity when trying to establish categories. Identity has no definition in the world of experience in general. We mean by '*identical*' that which is *similar* to the extent that we think relevant for the time being, more or less independently of its occurrence in space or time. All our categories are dependent on *subjective experience based on techniques*.

This implies that the process of defining categories is continually developing. Suppose we find that neuron, *A*, is in every way similar to neuron, *B*, until we examine it by a new test, we then redefine all those previously identical features as really belonging to two distinct categories. Since the whole process is open-ended, the sure observations on which further work can be erected are those where uniqueness or constant differences between neurons are obvious, and elsewhere we find numerous classes and categories in various stages of definition.

How to exploit identifiable neurons

The obvious advance is that we can put actual known neurons into our working models of how we think the nervous system functions. It is a far cry from the hypothetical or black box neuron inferred from physiological pathways.

The next few years will see the analysis of control of the neuron characters upon which identification is based. The form of the dendritic tree, the position of the cell body, the destination of the axon, the electrical properties of the membrane and the chemistry of the cytoplasm can be examined in a corresponding identified cell of another animal, or of the opposite side, following different treatments. There come to mind: the passage of time (development), learning, regeneration, hormone treatment, transplantation, stimulation, examination in mutants, comparison between species, and so on. All this awaits the neurobiologist.

Working with known cells means that analysis is at the level that corresponds in other ways to the animal's own functional units. Results will be related to mechanisms of differentiation of individual cells in a way that is impossible with mixed populations of cells. A way is now available for the molecular biologists to deal with identified cells one at a time.

The inadequacy of morphological studies

At one time there was a ray of hope that if one knew the total anatomical pattern of synapses between all the neuron types in a functionally distinct region of the nervous system, it should be possible to use the circuit to understand how the outputs of it are related to the inputs and to each other. Systems of synaptic regions amenable to analysis were sought, notably the early stages of the visual pathways in vertebrates and insects, and the supposedly simple nervous system of a nematode.

Primary visual fibres of vertebrates and arthropods are peculiar in having club-shaped terminals, like synaptic bags. Therefore behind the retina the regularly repeated distribution of the input is reasonably definable. One can search through all the anatomical information on all nervous systems (a large job) and not find so favourable a region for study by the combined methods of anatomy, physiology, and systems analysis. The insect optic lobe makes an analysis for colour, plane of polarisation, angle, direction of motion and possibly other properties of the whole visual field. Interneurons of the optic lobe can be recorded and synapses can be observed by electron microscopy. The stage appeared set for an elucidation of the circuitry behind the insect eye. However, it was demonstrated that about twelve different types of neurons occur in the region of the first synapses, and that they make at least a dozen recognisably distinct types of synapses with each other, in not obviously sensible combinations (Strausfeld, 1971). It is clear that the physiological interaction cannot be predicted from the morphological arrangement. Moreover, the structure of the lamina, where the first-order and related synapses are situated, is simpler in the fly than in other insects where the same structures have been examined. Even in the fly, however, the pattern of synapses, with the possibility of numerous closed loops of interaction between neurons in close proximity, inhibits conclusions about function, and certainly prevents a prediction of input/output relations. The anatomical framework must be filled out with time constants, latencies, phasic properties, synaptic gain, threshold and so on, even if basic synaptic properties such as inhibition and excitation could be distinguished. The only way to discover the physiological mechanism is to attack it directly with microelectrodes and cleverly designed stimulus situations. This forces us to accept the prime importance of recording physiological properties at every point, because information about what the neurons do is not available in any other way.

The nervous system of nematodes, rotifers etc., consisting totally of a few hundred identifiable neurons, may seem attractive as a simple morphological system but at present is intractible physiologically. Alike in simple nervous systems and in selected parts of larger examples, there is to be no escape from the integrated approach. Anatomy at all levels, microelectrode recordings from all neurons, model building and localised biochemical analyses are required on the one preparation. This certainly involves the identification of the neurons into as large a number of categories as can be consistently found, but it is not established that every neuron can be so recognised. Often we have only a physiological criterion as the basis of recognition.

However, it is no longer possible to accept an anatomical pattern of synapses as a guide to physiological pathways, if we accept that good morphological synapses can be non-functional (Mark *et al.* 1970; Mark, personal communication). We now have the exciting possibility that synapses can be switched off by trophic influences without changing their structure.

In conclusion, the anatomical techniques are extraordinarily *revealing* but the actual wiring diagram which might explain the function cannot be picked out from the mass of anatomical information. To make that step we have to record from identified cells, but not all neurons can be so identified. For many reasons therefore, physiological recording is paramount as a means of analysing what is going on, and this applies also to the analysis of formation of functional connections in studies of growth and regeneration.

Terminology

If we are looking for a general theory of how the nervous system is organised, the first pitfall is the explanation which is merely a restatement of the experimental results, wrapped up in an arbitrary terminology. Such explanations are readily detected, because they are in terms of invented attributes that are not observed by direct *analytical* techniques. The behaviour of a magnet, for example, is explained by its *attraction* to the magnetic north pole. To discover the subsystems that generate a more useful terminology one has to design experiments that pursue the subsystems into their own disciplines. Experiments by making transplants can show whether an axon's connections are determined already when transplanted, or if they are changed by the new location. The result carries no information about the choice of connections in terms of protein specificity, nature of gradients, mechanisms of territorial competition for sites, and so on. If the connection patterns are dependent upon protein specificity, then this cannot be inferred except from appropriate experiments. When the appropriate tests have been done, we see in the subsystems new categories from which a fresh terminology is born.

Wittgenstein referred to 'the bewitchment of our intelligence by language', and, as usual I don't understand his message. But I think I understand the danger of dreaming up a set of categories or entities, and then 'inferring' relations between them to 'explain' what is going on. Pavlov was responsible for about half of the terms commonly employed in studies of conditioning *because he carried out the defining experiments*. Only the experimentalist, working with a technique able to resolve the explanatory factors, can define a helpful terminology, and almost always he bewitches us into continuing to think for too long in the habit of that terminology.

The usage of some terms

Wiring diagram. The words suggest an approximation to an electrical equivalent. Therefore let us use 'wiring diagram' for the pattern of anatomical connections or presumed anatomical connections which produces the working diagram that fits with the functions observed physiologically. The difference from a connectivity pattern is that attention is paid to physiological pathways and non-functional anatomical contacts are omitted.

Connectivity pattern. The words suggest the whole pattern of connections. Therefore let us use 'connectivity pattern' for the total picture of morphological synapses. The difference from a wiring diagram is that nothing in a connectivity pattern is inferred except that observed structures are called synapses.

Identified neuron. A neuron which is individually unique and can be found reliably in every animal of that species. Identification is based on the whole anatomical appearance, branching pattern and axon destination. The intention is to label each neuron with an index number that is constant for the species.

Identified unit. A term based on microelectrode technique. A unit with fixed functions and physiological properties that can be found reliably in every animal, but which is not necessarily one particular identifiable anatomical neuron.

Physiological pathway. No more than what it says: when excitation at *A* causes excitation at *B*, observed in any way, a line can be drawn between *A* and *B* in a diagram. Physiological pathways are sometimes regarded as the mechanisms by which nervous systems operate.

References

Atwood, H. L. & Bittner, G. D. (1971). Matching of excitatory and inhibitory inputs to crustacean muscle fibres. *Journal of Neurophysiology*, **34**, 157–70.

Blakemore, C. & Cooper, G. F. (1970). Development of the brain depends on the visual environment. *Nature, London*, **228**, 477–8.

Bullock, T. H. & Horridge, G. A. (1965). *Structure and Function in the Nervous Systems of Invertebrates.* San Francisco: Freeman.

Davis, W. J. (1971). Functional significance of motoneuron size and soma position in swimmeret system of the lobster. *Journal of Neurophysiology*, **34**, 274–88.

Edds, M. V., Barkley, D. S. & Farnbrough, D. M. (1972). Genesis of neuronal patterns. *Neurosciences Research Program Bulletin*, **10**, 255–367.

Gaze, R. M., Chung, S. H. & Keating, M. J. (1972). Development of the retino-tectal projection in *Xenopus*. *Nature, New Biology, London*, **236**, 133–5.

Gaze, R. M., Keating, M. J., Székely, G. & Beazley, L. (1970). Binocular interaction in the formation of specific intertectal neuronal connections. *Proceedings of the Royal Society*, B, **175**, 107–47.

Govind, C. K., Atwood, H. L. & Lang, F. (1972). Differentiation of regenerating crab neuromuscular junctions. *American Zoologist*, **12**, 431.

Henneman, E., Somjen, G. & Carpenter, D. O. (1965*a*). Functional significance of cell size in spinal motoneurons. *Journal of Neurophysiology*, **28**, 560–80.

(1965*b*). Excitatory and inhibitibility of motoneurons of different sizes. *Journal of Neurophysiology*, **28**, 599–620.

Hirsch, H. B. B. & Spinelli, D. N. (1970). Visual experience modifies distribution of horizontally and vertically oriented receptive fields in cats. *Science*, **168**, 869–71.

Horridge, G. A. (1961). The organization of the primitive central nervous system as suggested by examples of inhibition and the structure of neuropile. In *Nervous Inhibitions*, ed. E. Florey, pp. 395–409. Oxford: Pergamon Press.

Horridge, G. A. (1968*a*). *Interneurons; their origin, action, specificity, growth and plasticity.* San Francisco: Freeman.

(1968*b*). Affinity of neurones in regeneration. *Nature, London*, **219**, 737–40.

Horridge, G. A. & Meinertzhagen, I. A. (1970). The accuracy of the patterns of connexions of the first- and second-order neurons of the visual system of *Calliphora*. *Proceedings of the Royal Society*, B, **175**, 69–82.

Jacobson, M. (1970). *Developmental Neurobiology*. New York: Holt, Rinehart & Winston.

(1971). Absence of adaptive modification in developing retinotectal connections in frogs after visual deprivation or disparate stimulation of the eyes. *Proceedings of the National Academy of Sciences, USA*, **68**, 528–32.

Mark, R. F., Marotte, L. R. & Johnstone, J. R. (1970). Reinnervated eye muscles do not respond to impulses in foreign nerves. *Science*, **170**, 193–4.

Meinertzhagen, I. A. (1972). Erroneous projection of retinula axons beneath a dislocation in the retinal equator of *Calliphora*. *Brain Research*, **41**, 39–49.

Mendell, L. M. & Henneman, E. (1971). Terminals of single *Ia* fibers. Location, density and distribution within a pool of 300 homonymous motoneurons. *Journal of Neurophysiology*, **34**, 171–87.

Peterson, R. P. (1966). Cell size and rate of protein synthesis in ventral horn neurones. *Science*, **153**, 1413–14.

Ryall, R. W., Piercey, M. F. & Polosa, C. (1971). Intersegmental and intra-segmental distribution of mutual inhibition of Renshaw cells. *Journal of Neurophysiology*, **34**, 700–7.

Strausfeld, N. J. (1971). The organization of the insect visual system (light microscopy). I. Projections and arrangements of neurons in the lamina ganglionaris of Diptera. *Zeitschrift für Zellforschung und Mikroskopische Anatomie*, **121**, 377–441.

Trujillo-Cenóz, O. & Melamed, J. (1970). Light and electron microscope study of one of the systems of centrifugal fibers found in the lamina of muscoid flies. *Zeitschrift für Zellforschung und Mikroskopische Anatomie*, **110**, 336–49.

Wall, P. D. & Egger, M. D. (1971). Formation of new connexions in adult rat brain after partial deafferentation. *Nature, London*, **232**, 542–5.

Young, D. (1972). Specific re-innervation of limbs transplanted between segments in the cockroach *Periplaneta americana*. *Journal of Experimental Biology*, **57**, 305–16.

Author Index

Figures in bold type indicate pages on which references are to be found.

Author Index

Author Index

Subject Index

Subject Index